Soil Conservation and Management in the Humid Tropics

This volume is based on the Proceedings of the International Conference on Soil Conservation and Management in the Humid Tropics held in Ibadan, Nigeria, in June 1975.

The Conference was jointly sponsored by the International Institute of Tropical Agriculture and the Agricultural Research Council of Nigeria.

Soil Conservation and Management in the Humid Tropics

Edited by

D. J. Greenland
Professor of Soil Science, University of Reading, England, and formerly Director of Research, International Institute of Tropical Agriculture, Ibadan, Nigeria

and

R. Lal
International Institute of Tropical Agriculture, Ibadan, Nigeria

A Wiley–Interscience Publication

JOHN WILEY & SONS
Chichester · New York · Brisbane · Toronto

Library of Congress Cataloging in Publication Data:

International Conference on Soil Conservation and Management in the Humid Tropics, Ibadan, Nigeria, 1975.
Soil conservation and management in the humid tropics.

Papers from the conference sponsored by the International Institute of Tropical Agriculture and the Agricultural Research Council of Nigeria.

"A Wiley-Interscience publication."
"Bibliography on soil erosion research in Africa": p.
Includes index.

1. Soil conservation—Tropics—Congresses. 2. Soil erosion—Tropics—Congresses. 3. Agriculture—Tropics—Congresses. I. Greenland, D. J. II. Lal, R. III. International Institute of Tropical Agriculture. IV. Nigeria. Agricultural Research Council. V. Title.

S625.T76157 1975 631.4'913 76-56838

ISBN 0 471 99473 1

Printed in Great Britain by
William Clowes & Sons, Limited
London, Beccles and Colchester

Contributors

N. Ahmad	*Department of Soil Science, The University of the West Indies, St Augustine, Trinidad.*
P. M. Ahn	*Department of Soil Science, University of Nairobi, Kenya.*
M. De Boodt	*Department of Soil Physics, State University of Ghent, Coupure Links* 533, *Belgium.*
R. G. Dumsday	*Department of Agricultural Economics, La Trobe University, Melbourne, Australia.*
K. A. Edwards	*EAAFRO, P.O. Box 30148, Nairobi, Kenya.*
A. O. Falusi	*Nigerian Institute of Social and Economic Research, Ibadan, Nigeria.*
R. Fauck	*ORSTOM, Paris 75008, France.*
J. C. Flinn	*International Institute of Tropical Agriculture, Ibadan, Nigeria.*
D. Gabriels	*Department of Soil Physics, State University of Ghent, Coupure Links* 533, *Belgium.*
W. R. Gardner	*Department of Soil Science, University of Wisconsin, USA.*
D. J. Greenland	*International Institute of Tropical Agriculture, Ibadan, Nigeria.*
N. W. Hudson	*National College of Agricultural Engineering, Silsoe, Bedfordshire, UK.*
J. M. Kalms	*ORSTOM, Bouake, Ivory Coast.*
A. H. Kassam	*ICRISAT, Hyderabad-*50016, *India.*
J. M. Kowal	*Department of Botany, University of Liverpool, UK.*
R. Lal	*International Institute of Tropical Agriculture, Ibadan, Nigeria.*
J. Lenvain	*Department of Soil Physics, State University of Ghent, Coupure Links* 533, *Belgium.*
L. Maene	*Department of Soil Physics, State University of Ghent, Coupure Links* 533, *Belgium.*
W. C. Moldenhauer	*USDA-ARS North Central Soil Conservation Research Centre, Morris, Minnesota, USA.*

S. O. Olayide	*Department of Agricultural Economics and Extension, University of Ibadan, Nigeria.*
B. N. Okigbo	*International Institute of Tropical Agriculture, Ibadan, Nigeria.*
C. A. Onstad	*USDA-ARS North Central Soil Conservation Research Centre, Morris, Minnesota, USA.*
C. R. Panabokke	*ARI Gannoruwa, Peradeniya, Sri Lanka*
I. Pla	*Instituto de Edafologia, Facultad de Agronomico, Maracay, Venezuela.*
P. Ponsana	*Central Region Agricultural Centre, Chainat, Thailand.*
E. Roose	*ORSTOM, Abidjan, Ivory Coast.*
S. A. El-Swaify	*College of Agriculture, University of Hawaii, Hawaii.*
S. Tienseemuang	*Central Region Agricultural Centre, Chainat, Thailand.*

Contents

Foreword

PROFESSOR E. W. RUSSELL

Soil conservation has been generally accepted as an essential principle in all good farm management systems since the early 1930's when H. H. Bennett in America persuaded the US Government that it had a duty to help farmers to farm their lands in such a way that they did not deteriorate with use. This principle has long been accepted in Western Europe, where improved systems of farming were being actively developed in the seventeenth century, and it was embodied in such precepts as 'a man should always aim to hand over his farm to his son in at least as good a condition as he inherited it from his father'.

The improved systems developed in Western Europe were all conservative—that is they conserved or improved the fertility of the soil—certainly up to the onset of the tractor. But these systems did not need to pay much attention to soil erosion, for the rainfall pattern in Western Europe is not one that causes serious erosion, so that farmers brought up in these regions were unaware that many of their practices were only conservative because of the relative rarity of high-intensity rain storms. Lack of appreciation of this inherent limitation had very serious consequences when emigrant farmers from Western Europe began settling in areas where high intensity rain storms were an appreciable hazard, with the consequence that soil erosion often became widespread in many of the areas they settled in.

Indigenous systems of farming developed in the tropics and subtropics, where heavy to very heavy rain storms are characteristic of the climate, are also usually conservative, and result in little soil erosion. Otherwise they would be unstable and would have died out. But these systems are nearly all characterized by low to very low crop yields, and usually only a small proportion of the land in an area is under crop at any one time. Thus these systems are only well adapted to low population densities and a standard of living that is not far removed from starvation in unfavourable seasons.

Increasing population densities are producing an increase in the proportion of land under cultivation in any one year, either by shortening the time the land is under a fallow rest or by cultivating land that is traditionally considered unsuitable for cultivation, without any other modification in the system. This has resulted, in a number of areas, in the development of a serious soil erosion problem due to this breakdown in the equilibrium between population density and the traditional system of farming. Further, in many of these areas simple and acceptable changes in the land system, made in the hope of achieving a higher productivity from the land, have resulted in serious soil erosion since they were not developed in the context of a foreseeable but preventable erosion hazard.

The cause of soil erosion is always an increase in the amount of storm water running off the soil surface, for it is this water that disperses and carries away the soil particles. But even if a system of land use is developed which reduces the amount of soil being carried away by the water, the loss of water itself can be very serious in regions where crop yields are limited by inadequate seasonal rainfall for two reasons: it causes an actual loss of water and it reduces the depth of soil wetted by a storm, and this shallowing of the depth of wetted soil also increases loss of crop growth during breaks in the rain.

Soil erosion and water run-off are usually two overriding factors limiting the kinds of farming systems that can be introduced into regions subjected to heavy rain storms. Since the indigenous systems developed in the areas are now largely inappropriate, due to pressure on the land for increased food production, one of the most urgent problems now facing agricultural scientists is the development of a new, appropriate and conservative system of agriculture, adapted to the local farmers and the region, that will allow greatly increased yields. This must involve on the one hand continued studies on the factors causing erosion and run-off, but also of the various practices that can reduce the liability of erosion and increase the rate of infiltration of water through the soil surface to below the depth of rooting of the crop. This work must involve a very wide range of agricultural and scientific expertise.

It is most gratifying to note how actively these studies are being pursued in the developing world at the present time, and it is appropriate and important that summaries of the present state of knowledge be provided from time to time to allow both the scientist and the agronomist to survey the field and identify those areas most in need of further study if new productive and conservative systems of agriculture are to be developed in these very extensive and important areas. This volume, by providing such a comprehensive review of the present state of knowledge in this field, should fill an important niche in the literature on the subject. I am sure it will receive a warm welcome from everybody who is concerned with raising the level of agricultural production and hence the level of nutrition of the large proportion of the world's population which lives in lands liable to soil erosion if suitable precautions are not taken to prevent it occurring.

Professor Emeritus of the Department
of Soil Science,
University of Reading

PART 1

Introduction

1.1

The Magnitude and Importance of the Problem

D. J. GREENLAND

Soil erosion is a natural and worldwide phenomenon. Whenever the weathering of rock produces the fine-grained material of soil, wind and water will tend to wash the soil away, depositing it wherever the movement of wind or water ceases. In the humid regions of the world, an equilibrium is normally established between the slow process of soil formation and removal of soil by erosion. This is because the erosive process, except on very steep slopes, is correspondingly slow provided that the soil is covered by vegetation. Problems arise when the vegetation is removed, by grazing animals, or by agriculturists in preparation for planting, or by engineers undertaking construction work. The bare soil is exposed to the direct action of rain, and the impact of raindrops causes detachment of soil particles which are then carried away in water moving over the soil surface. Any soil covering will reduce the impact of the raindrops and so prevent detachment, but vegetation has the added advantage that it is rooted in the soil and the roots help to bind the soil in place. Further, the organic matter added to the soil as litter supports the animals and microorganisms which bind the smaller soil particles into aggregates, again hindering their detachment.

Soil erosion is not simply a function of soil exposure. It depends as well on inherent properties of the soil, the landscape in which it occurs, and the climate, all of which affect the extent to which rain can enter the soil, and thus the proportion which washes over the soil rather than seeping into it. One of the most important reasons why the visible effects of soil erosion are so much more pronounced in the humid tropics is the very high erosivity of the rainfall. The intensity of tropical rainstorms is much greater than that of the rainfall in more temperate areas. While the intensity of tropical storms has been widely recognized, it is only recently that it has been established that the erosivity of rain in the humid tropics is substantially greater than, for instance, in the south and midwest of the United States, from where most information on erosion control measures has come. Data relating to the erosivity of tropical rain are given in Parts 2 and 3. Because of this greater erosivity, and also because of the very different social and economic conditions prevailing in the humid tropics, it is important that correct data are assembled on which to base appropriate soil management systems. Important recent publications (Rapp *et al.*, 1972; Jones and Wild, 1975) have assembled much information relating to erosion in the subhumid to semi-arid tropics. Much remains to be done in the humid tropics.

At this stage it is well to consider the definition of humid tropics. For present purposes 'humid' will be taken to refer to those areas where the rainfall exceeds evaporation for at least seven and a half months in the year. Although the tropics are defined as the areas lying between the latitudes of the Tropic of Cancer and the Tropic of Capricorn (23°N and 23°S), the humid tropics according to the above definition roughly conform to those areas between the latitudes of 10°N and S (Figure 1.1). Evaporation in these regions is of the order of 100 to 150 mm per month, so that the rainfall must normally exceed 1000 mm a year. Often it will be substantially greater than this. The dry season, by definition, cannot exceed four and a half months. This implies that the natural vegetation in most of these areas will be forest, although man's activities may well lead to its replacement by grassland and this has happened in the less humid parts of the region, such as the Guinean (derived) Savannah Zone of West Africa.

Wind erosion does not occur when the soil is wet, or when there is an adequate cover of vegetation or a high frequency of trees in the landscape. Thus it is not generally a problem of any significance in the humid tropics, although in areas of derived savannah, blowing away of fertile topsoil may occur if the vegetation is removed by burning or overgrazing and the soil exposed during the dry season.

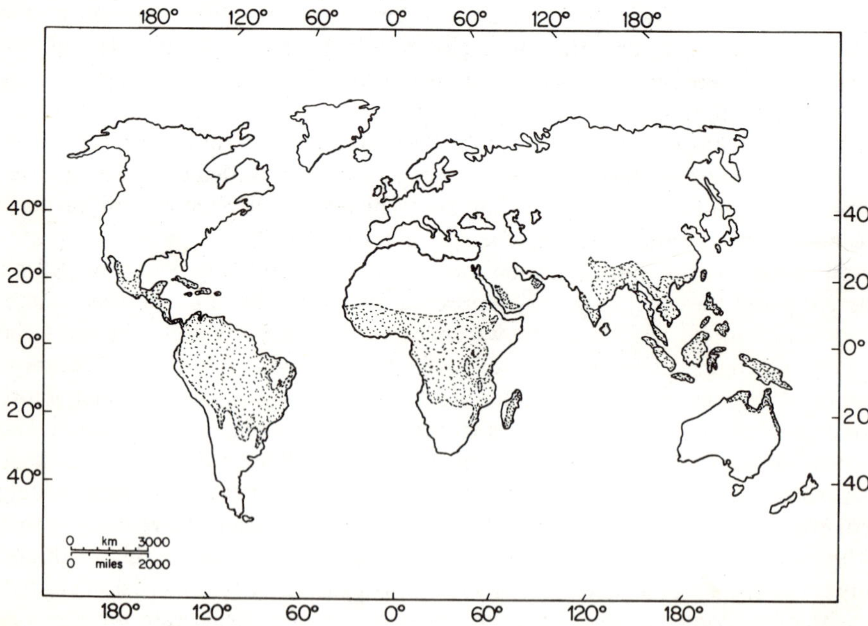

Figure 1.1. Areas of the tropics with five or more humid months. Derived from Map 5, 'Seasonal climates of the earth', in Landsberg *et al.* (1963)

Plate 1. Deep gullies in the Enugu region of Eastern Nigeria, mostly initiated in the past 50 years with increasing intensity of land use

[*Facing page* 4

Plate 2. Exposure of infertile subsoil due to sheet erosion results in a scanty vegetation cover

Plate 3. Loss of surface soil and plant nutrients results in poor or no crop growth

Plate 4. Burial of surface soil and crops by sedimentation of eroded soil (courtesy Dr W. Moldenhauer)

Plate 5. Example of effective soil and water conservation systems; a permanently vegetated back slope on a parallel terrace system, developed in the US (courtesy Dr W. Moldenhauer)

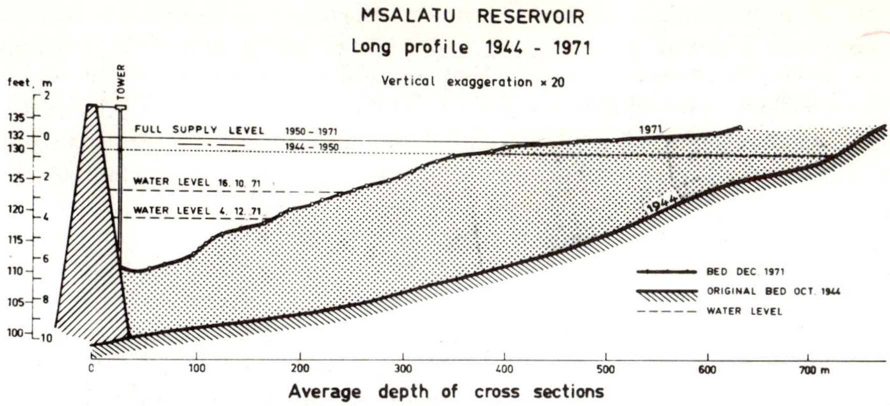

Figure 1.2. Effects of eroded soil on reduction of water storage, Msalatu reservoir, Tanzania. Reproduced, with permission, from Rapp *et al.* (1972) (p. 297)

The occurrence of undesirable soil erosion in the humid tropics is thus largely associated with the removal of soil, after it has been cleared of vegetation, by intense rainfall. Wherever agriculture is practised, wherever roads and paths are cleared of vegetation, excessive removal of soil is likely to occur. The removal of soil may result in the formation of deep gullies which may link and join and devastate a whole landscape (Plate 1); it may remove the surface soil, without causing gullies to develop, and expose a bare and infertile subsoil on which it is difficult for vegetation or crops to develop (Plate 2); and invariably it leads to some loss of the richest part of the soil and carries away plant nutrients and organic matter, reducing to a greater or lesser extent the ability of that soil to produce crops (Plate 3).

The effects of soil erosion are by no means confined to those due to the actual loss of soil from the site where it originated. Effects arise due to the reduced permeability of the newly exposed soil, and the reduced water retention; the increased amount of water running off the surface, leading ultimately to floods, changes in depths of river beds, erosion of stream and river banks, and related effects. The cutting of gullies lowers the water-table in the surrounding area, so causing the dry periods to have a more deleterious effect on vegetation. The carriage of soil particles in the runoff water leads to problems where the water stops moving and the silt is deposited. This may be simple burial of crops or more fertile topsoil (Plate 4); it may lead to deposition in lakes and reservoirs, which become silt traps rather than water storages (Figure 1.2); and almost always it leads to a reduction in the permeability of soil where it is deposited, as the fine particles settle into water-conducting pores of the soil profile.

Although these effects of soil erosion have been fully described in publications such as *Vanishing Lands* by Jacks and Whyte (1972) and the classical volume on the subject, Bennett's *Soil Conservation* (1939), the problem remains one of the most serious threatening the future of mankind.

In some areas much has been learnt and much has been done, but this has mostly been confined to the richer, agriculturally advanced areas of the temperate zone (Plate 5). Today the most rapid increase in the world's population is occurring in the underdeveloped countries of the tropics. The soils of these areas are generally much less productive than they might be, and if more intensive use leads to further loss of fertility and further erosion, the present potential to feed the burgeoning population of these regions will be lost.

It is quite often stated that the technical solutions to the soil erosion problem are well known, and that it only remains to apply these known solutions. Although it is true that given the necessary financial resources, the required trained technicians, the political strength and the necessary confidence to enforce land management policies on farmers and other landowners, much can be done to control soil erosion, details of designs for erosion control may be less than perfect because of large gaps in our knowledge of erosivity of climates and erodibility of soils. Nevertheless, erosion control programmes giving a degree of control can be formulated for most parts of the world. Handbooks of erosion control measures are available (USDA, 1965; FAO, 1965), and the Food and Agriculture Organization of the United Nations (FAO, 1971) has produced a bulletin relating to the legislative measures needed to enforce proper control of soil erosion.

Unfortunately, the basic requirements for establishing erosion control by such measures are seldom met in the humid tropics. Most governments have demands to meet apparently more pressing than the establishment of soil erosion control, they are extremely short of soil scientists and engineers who are able to develop the necessary erosion control and land-use policies and apply them, they lack the political and social systems which are necessary to enforce control, and the land pressures and need for increased food by the population tend to override any considerations which would appear to reduce, even if temporarily, the ability to produce more food cheaply by cultivating more land.

It is thus more correct to state that some technical solutions are known, but technical solutions which can be applied, given the existing social and economic constraints, have still to be developed. Most of the textbooks present designs for land and water management (e.g. FAO, 1972) requiring considerable use of heavy equipment. As with many other research station 'developments' in agricultural science, they are meaningless as far as the bulk of farmers in the humid tropics are concerned. Thus in much of Africa and Latin America soil erosion is becoming more serious and more widespread as the small farmers are forced by population pressure to cultivate land more frequently and more intensively, although a few relatively well-developed and sophisticated schemes may be held as a general example which others are expected to follow.

It will be seen in the chapters which follow that our understanding of the causes and prevention of soil erosion is improving, so that we are passing from an era where unrealistic remedies were often discussed. The failure to produce remedies which meet the social and economic constraints has meant that erosion has continued to flourish in much of the humid tropics. This continuing degradation of the world's most important resource is also described, in the hope that wider appreciation of the problem will spur those concerned to greater efforts in preventing further damage.

REFERENCES

Bennett, H. H., 1939, *Soil Conservation*, McGraw-Hill, New York.

FAO, 1965, Agric. Paper No. 81.

FAO, 1971, Legislative principles in soil conservation, Soils Bull. 15.

FAO, 1972, Environmental aspects of natural resources management, Agriculture and soils, Agric. Service Bull. No. 14.

Jacks, G. V., and R. O. Whyte, 1972, *Vanishing Lands*, Arno.

Jones, M. J., and A. Wild, 1975, Soils of the West African Savanna, Tech. Comm 55, CAB, Farnham Royal.

Landsberg, H. E., *et al.*, 1963, *Weltkarten zur Klimakunde*, Springer-Verlag, Heidelberg.

Rapp, A., L. Berry and P. Temple, 1972, Erosion and sedimentation in Tanzania, *Geografiska Annaler*, **54 A**, Nos. 3 and 4.

USDA, 1965, Agric. Handbook No. 282.

Wischmeier, W. H., and D. D. Smith, 1965, Predicting rainfall erosion losses from crop land, Agricultural Research Service, USDA, Agric. Handbook No. 282.

PART 2

Basic Aspects of Soil and Water Conservation

2.1

The Factors Determining the Extent of Soil Erosion

N. W. HUDSON

2.1.1. ANALYSIS OF THE PROBLEM

In the battle against soil erosion we are losing on every front, and what we need are practical control measures. The primary objective of erosion research must be to obtain the information required to design or develop appropriate measures. It has become customary to refer to 'soil and water conservation' or 'soil and water management'. It would be better to have one single word which includes both soil and water so that they are thought of together, because one can seldom be changed without affecting the other. However, for the present, only soil erosion will be discussed, except when the problems of water conservation and management arise as part of soil conservation. Also, although there are some situations in the humid tropics where wind erosion is a problem, they are few, and attention is given only to water erosion. Finally, as the problems of erosion on arable land are more pressing than those on non-arable land, arable lands will be discussed more fully.

2.1.2. SOIL EROSION ON ARABLE LAND BY WATER

The factors which affect the amount of erosion which will occur in given circumstances are illustrated in Figure 2.1. Erosion arises from the effects of rainfall on the soil, and is determined by:

(i) *Erosivity*—specifically and solely a property of rainfall, which can be quantitatively evaluated as the potential capacity of rain to cause erosion in given circumstances.

(ii) *Erodibility*—specifically and solely a property of the soil, which can be quantitatively evaluated as the vulnerability of soil to erosion in given circumstances.

(iii) *Management*—a wide term covering all the factors directly under man's control, such as the choice of land use, choice of crop, method of crop

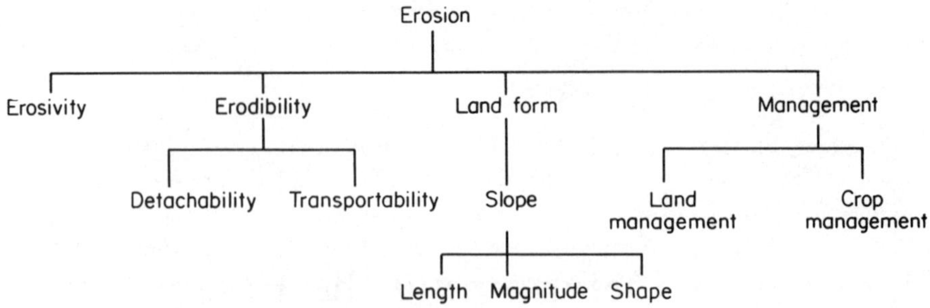

Figure 2.1. Factors contributing to erosion

production, down to details (but highly important details) like plant population and manurial practices. Each of these must be considered in more detail.

(iv) *Land form*—which includes the length and steepness of slopes, and their shape (concavity or convexity) and uniformity of shape. Usually for agricultural use land areas to be cultivated tend to be of finite length and moderate slope, but this is not always true, particularly in formerly forested areas.

2.1.2.1. Erosivity

The general principle is known, that erosivity is closely related to intensity, and can be evaluated by calculations based on kinetic energy. Up to the present, several indices described on pp. 70 and 71, EI_{30} (Wischmeier and Smith, 1960, $KE > 1$ (Hudson, 1976), AI_m (Lal, 1976) or for regional studies p^2/P (Fournier, 1960), have been used. The limitations of these rainfall erosivity indices are discussed in more detail in Parts 3 and 4. Some countries have calculated long-term average erosivity values for different centres (i.e. variation in space) or weekly values during the rainy season (i.e. variation in time). Where this has not been done, a common difficulty is lack of suitable data. Sometimes the raw data (i.e. charts from recording raingauges) are available but have not been processed into a form suitable for computation of erosivity. It is therefore very necessary that these data be collected on an international basis. Further, in almost all tropical climates we need more studies of rainfall. In the past it was considered sufficient to record amounts of rain. For the purpose of erosion control we need to know about erosivity, maximum intensities, intensity–duration relationships, and frequencies of severe storms.

2.1.2.2. Erodibility

The search for a quantitative measure of resistance to erosion, which can be calculated from measurable soil properties, is as old as soil erosion research. The present state of the art is discussed in more detail in later chapters, but some general points can be made here. The difficulties are well known. Even on one soil type erodibility varies considerably from pedon to pedon in the field; the problem of whether to measure erodibility in the field or on so-called undisturbed samples in the laboratory is not readily resolved. We still need to know how to measure cohesive forces adequately, and how to quantify soil structures. Since we have no general solution, a rigorous solution for particular cases cannot be expected. However, qualitative assessments are often readily made, such as, for example, two soils are known to have nearly the same erodibility, while another is observed to have much more resistance. Before we dismiss such judgments too lightly, let us remember how the Soil Conservation Service of the US Department of Agriculture started with very few data on erodibility (in fact seven values), then extrapolated this to 23 main soil types, and subsequently used subjective assessment to extend to hundreds of soil series (Wischmeier and Smith, 1960).

What needs to be done in the study of erodibility? Any solution is almost certain to be empirical. All empirical relationships are valid only within the conditions under which they were generated, and before being applied in other situations they must be validated for the new situation. There are thus three things to be done to establish values for the humid tropics: (i) examine and refine the existing general procedure for use in the humid tropics; (ii) test its validity; and (iii) apply it to different soils of the region.

2.1.2.3. Management

The principles are well known, as they are the same as for other climatic zones: the importance of mulch or crop cover, the need for controlled flow of surface water, the association between efficient farming and good erosion control, and many others (Hudson, 1976). However, because of the greater erosivity of the climate in the humid tropics, the maintenance of a cover over the soil assumes greater importance. The management practices for some situations are well established, but in other cases the answers are less clear. Intensification, development, and mechanization bring new erosion hazards, for which the answers are not always available.

More local studies of different management practices are required. Even when the answers can be predicted, local experiments are justified for their demonstration value.

2.1.3. LAND FORMS

Both degree and slope length are important factors influencing soil erosion. Erosion generally increases exponentially with increase in slope; the exponent for tropical soils approaches 2, though it has been reported to vary between 1·3 and

2·1. Length of slope has a similar effect on soil loss as degree of slope. The build-up of the amount of surface runoff, its velocity and depth, on longer slopes increases erosion hazard. However, the relation between slope length and erosion is complicated by the shape of slope, viz., whether convex, concave, complex or regular. In the case of irregular slope, it is the steepest section of slope which may dominate the erosion.

2.1.4. CONTROL OF EROSION ON ARABLE LANDS

Of the factors influencing erosion, we can do nothing about erosivity or land form and nothing significant about erodibility, and so control must be achieved through management.

In general, the methods of erosion control are all known. It is unlikely that a new basis for control such as the concept of bench terracing will be found, although new herbicides have revolutionized the importance of zero tillage methods (Part 4). In the regional case, further knowledge and experience of the application of known erosion control measures must be sought continuously.

Land-use classification is the first vital step towards erosion control. This is one of the few cases where the admirable work of the US Soil Conservation Service is not applicable to the humid tropics. The conventional Land Capability Classification is excellent for moderate slopes and low land-use pressure, but unsuitable for the steep slopes and severe land hunger found in large parts of the humid tropics. However, other methods have been developed from experience in Taiwan, Malaysia and the Caribbean countries (Sheng, 1971). This work needs to be more widely applied to the other parts of the humid tropics, supplemented as far as possible by local data on erosivity and erodibility.

Agronomic and biological factors in control are discussed in later section by Lal, Roose, and Edwards. Again the principles are known, but all local farming systems need to be studied in relation to their own specific needs. There are so many variables that it is not possible to evaluate their separate effects, no matter how sophisticated the experimental design. Going back to the basic objectives of erosion research, we are looking for design criteria; so practical answers to practical problems are more important than measuring the effect of minor variations within the farming system.

Mechanical control is discussed by Moldenhauer and Fauck in Parts 4 and 5. Both the principles and the details of different designs for land management are well known, but data on rainfall erosivity and soil erodibility are needed in order to arrive at design specifications for the tropics. Simple 'look–see' trials of established techniques would seem to be preferable to repeating laborious studies such as spacing, or the effect of slope. The economic viability of different methods also requires much more attention, as do the social factors involved in getting widespread acceptance of suitable methods. Some aspects of these factors are discussed in Part 4.

2.1.5. EROSION ON NON-ARABLE LAND

2.1.5.1. Grazing Land

In general, the erosion problems of tropical grasslands are understood and the solutions are known (Manshard, 1968). When there is serious erosion from grazing land the cause is almost certain to be poor management (such as overgrazing) and the correction will be improved management. However, it is not always easy to implement the technical solution to the problem (such as destocking) because of less amenable social or economic problems. Social customs and land tenure are usually at the root of the overgrazing problems and it is towards the solution of these that any research should be directed.

2.1.5.2. Non-food Crops and Tree Crops

When non-food crops are grown on plantations, the erosion is inversely proportional to the level of management. With good management there are high inputs of technology and of investment, known practices for control of runoff are adopted, and erosion is under control. The danger arises when high standards of land management are not maintained. Production of tobacco, tea, coffee, sisal, oil palm, and rubber can all lead to serious erosion when not accompanied by appropriate erosion control measures.

This situation is likely to occur in the case of exploitive management which chooses not to invest in erosion control, or when such crops are grown by smallholders who lack the resources to invest in erosion control, or where adequate advisory services are not available.

The time of worst vulnerability is usually during the establishment period, when the original forest cover has been removed but the replacement is not yet providing protection. In the case of long-duration crops like oil palm or tea, there is some justification for tolerating high rates of erosion during the establishment if this is compensated by little or no erosion during the rest of the time the plantation is in production.

2.1.5.3. Reclamation of Eroded Areas

Badly eroded areas, such as badlands, scalds, and gullies, are usually local manifestations of local problems. There may be a need for local investigations, but on the whole the mechanisms are understood (Hudson, 1976) and control comes down to an analysis of whether reclamation is economically justified. There is usually little need for a research programme, but plenty of scope for local trials of techniques which have been found to be useful elsewhere.

REFERENCES

Fournier, F., 1960, *Climat et Erosion*, Presses Universitaires de France, Paris.
Hudson, N. W., 1976, *Soil Conservation*, Batsford.
Lal, R., 1976, Soil Erosion on Alfisols in Western Nigeria, III. Effect of Rainfall Characteristics, Geoderma, **16**, 389–402.
Manshard, W., 1968, *Tropical Agriculture* (trans. D. A. M. Naylon), Longman, London.
Sheng, E. C., 1971, A treatment oriented land capability classification scheme (for hill marginal lands in the humid tropics), Latin American Watershed Management Seminar, La Plata, Argentina, 1971.
Wischmeier, W. H., and D. D. Smith, 1960, A universal soil-less estimating equation to guide conservation farm planning, *Trans. 7th Intern. Cong. Soil Sci.*, **1**, 418–25.

2.2

Soil Structure and Erosion Hazard

D. J. GREENLAND

2.2.1. INTRODUCTION

This chapter is concerned with soil erosion due to the effects of rainfall. It is not concerned with wind erosion. Erosion due to water arises when rainfall does not all penetrate the soil, and the excess runs off the land, carrying with it detached soil particles. The detached material is usually the topsoil, where plant nutrients are most heavily concentrated. The consequent exposure of subsoil often leads to a lower rate of water entry, increased runoff and further soil loss. Thus a continuing cycle of soil deterioration develops. Soil structure is important to soil erosion because it largely determines the rate at which water can enter the soil, as well as the resistance of soil particles to detachment by rainfall impact and subsequent removal in the surface runoff.

Marshall (1962) defines soil structure as 'the arrangement of the soil particles, and of the pore space between them. It includes the size, shape and arrangement of the aggregates formed when primary particles are clustered together into larger, separable units'. The structure of surface soils is usually given most attention in relation to soil erosion, because it is most subject to deterioration under raindrop impact, and due to agricultural practices its condition can be readily altered. However, subsoil structure is also important, as infiltration in a water-saturated profile is determined by the horizon of lowest conductivity, and that is often a subsoil horizon of higher clay content than the surface soil.

The stability of any given structural organization is particularly important in relation to erodibility. This arises firstly in relation to the ease of detachment of particles from aggregates. Smaller particles are more readily transported by moving water than larger ones, and hence if aggregates are stable against disruptive forces, little erosion will arise, provided the aggregates are sufficiently large, or themselves cohere to each other sufficiently strongly to prevent movement. The stability of the aggregates is also important insofar as detached particles of clay and silt are likely to be washed into the coarser pores of the existing structure, and so cause a decrease in conductivity of the soil for water. The stability of aggregates of the immediate surface soil is particularly important, since it is these aggregates which are exposed to the direct impact of raindrops. If they break down into their constituent particles under this impact, a surface seal can be readily formed, reducing infiltration and increasing runoff, and leaving an exposed layer of small, readily transported particles.

Many of the older textbooks on tropical soils state that the old, red, leached, acid soils of the tropics are structurally very stable, and have good permeability, so that they have few physical problems. This generalization is much too broad. Many have weak structures, and the physical properties of tropical soils need to be considered rather critically in relation to their constitution.

2.2.2. POROSITY AND INFILTRATION

Adequate descriptions of the 'structure' of tropical soils are very few and measurements of porosity which are useful in relation to understanding of water conductivity even fewer, although pore-size distributions of non-swelling soils can be determined quite adequately from water content–suction curves (pF curves), using the Kelvin equation

$$r = \frac{2\gamma \cos \theta}{\rho g h}$$

(where r is the 'equivalent' pore radius, γ the surface tension of water, θ the contact angle of water on soil, ρ the density of water, g the gravitational force and h the suction, expressed as height of a water column). The larger transmission pores are all that are of interest in relation to water movement and infiltration. These are drained at suctions less than 100 mbar. Removing soil samples from the field normally leads to breakdown of pores of this size, approximately those larger than 30 μm, and so determinations must be made on large, intact cores, or *in situ* in the field, which may be the reason for the lack of data relating to this porosity range.

Visual assessments usually indicate that coarse porosities are high in surface soils under natural vegetation, where active faunal populations create many channels by burrowing activities. When many worms are present, the casts they create form a very coarsely aggregated structure, with many large channels between the casts. P. O. Aina, working with Dr. G. Wilkinson at Ife in Nigeria, obtained infiltration data for soil under tropical forest fallow and the very high rates of water entry show clearly the importance of this natural, faunal-created structure (Table 2.1). Their 'infiltration capacity' refers to water entry into saturated soil when lateral flow is not restricted, and 'infiltration rate' to the rate of entry when the horizontal movement is prevented. When subsoil permeability determines infiltration *rate*, clearing and cultivating has little effect, but compaction of the well-structured topsoil on clearing causes the infiltration *capacity* to fall by about half. In the Oba soil the decrease in infiltration rate in the second year of cropping is presumably due to washing of fine particles from the surface soil into coarse pores below. Lal (1975) at Ibadan, Trouse and Baver

Table 2.1 Equilibrium infiltration capacities and rates of water entry into crust-free Iwo and Oba soils in bush fallow and after two subsequent years of annual crops (maize) (Wilkinson and Ania, 1976)

	Iwo		Oba	
Soil condition	Infiltration capacity (cm/hr)	Infiltration rate (cm/hr)	Infiltration capacity (cm/hr)	Infiltration rate (cm/hr)
Bush fallow	118·9	21·1	107·7	25·4
After first cropping year	44·7	20·5	48·8	25·1
After second cropping year	49·5	22·4	21·6	9·1

(1965) in Hawaii, Sanchez (1974) in the Amazon region of Peru, and van der Weert (1974) in Trinidad have obtained similar results.

Although the infiltration capacities of many tropical soils may change from over 100 cm/hr to less than 10 cm/hr, the latter rate is still high by comparison with most temperate zone soils, or with many heavy clays. This reflects the stability of the 'microaggregates', which will be discussed further below. The rainfall intensities to which these soils are subject often exceed 10 cm/hr, so that some slight runoff occurs even under tropical rain forest. Because of the frequent restriction lower in the profile, and high surface soil conductivity, lateral water movement near the surface is a frequent phenomenon in these areas.

The data of Aina and Wilkinson refer to a sandy soil with only 10 per cent clay. Higher clay contents give rise to lower infiltration rates, particularly after clearing and cultivating, and substantial runoff is normally measured in such conditions during heavy storms. Trouse and Baver, for instance, report for the 'humic latosol' they studied an infiltration rate of 8 cm/hr without traffic, and 2 cm/hr after compaction of the surface soil due to harvesting of sugar cane. Lal and Mensah-Bonsu (1975) report hydraulic conductivities of surface soils (Egbeda) with 10 per cent clay of 7·5 cm/hr, but subsoil values, where the clay content was 30 per cent of only 0·072 cm/hr.

2.2.3. FORCES LEADING TO STRUCTURAL CHANGES

(a) Improvement. Forces tending to increase the porosity of soils are: (i) Physical, associated with shrinking and swelling, in turn due to wetting and drying or freezing and thawing. Obviously, these are unimportant in those tropical soils where the swelling potential is very little, and freezing does not occur. The term 'structureless' is sometimes used to describe the deep subsoils of tropical red earths—'apedal' is perhaps a better word. The lack of obvious structural units is associated with the absence of conditions which can create them. (ii) Biological, associated with the action of the soil fauna and plant roots. The fauna are most important, and the activities of worms, ants, and some termites in particular are responsible for the open-surface structures of tropical forest soils. The activities of the fauna are of course dependent on an adequate food supply, and the reduction in this supply on clearing and cultivating leads to a drastic reduction in populations. The earthworm population under secondary forest was found to fall at IITA by a factor of 10 in each of the first two years after clearing (COPR, 1974 and 1975). (iii) Tillage: appropriate manipulation of the soil with hoes or more advanced machinery can create increased porosity.

(b) Deterioration. Forces tending to reduce porosity include: (i) Raindrop action (Cohron, 1971). Data for the kinetic energy associated with rainstorms of different intensity are given in Chapter 3.1. This may exceed 10^5 Nm^{-2}. An average storm might be 10 per cent of this. (ii) Treading and 'wheeling' by animals, man and machinery. These forces are of similar order of magnitude, although tractor wheels can exert greater forces if wheel slip occurs, and sheer forces exerted by some implements may be higher. (iii) Rapid swelling, and air pressures, associated with rapid wetting of previously dry soils, important only in soils of medium to heavy texture.

2.2.4. STABILITY OF STRUCTURE

In considering the relationship between soil structure and erosion hazard it is important to know the ability of a soil to withstand the forces leading to deterioration of structure. The coarse pores which are of most interest are normally interaggregate pores, and because of this most work has been concerned with determining aggregate stability, since if the aggregate remains intact, the interaggregate pores will also persist. What then determines aggregate stability?

To answer this question it is necessary to use a concept of aggregate structure. Soil aggregates are formed by the association of clay particles into domains, of domains and silt particles into microaggregates, and of microaggregates and sand particles into aggregates. Size limits for these units which may be suggested are:

Domains up to 5 μm
Microaggregates 5–1000 μm
Aggregates 1000–5000 μm
(1–5 mm) (Greenland, 1971).

The forces which hold domains together are electrical, and usually relatively large because of the very close approach of the particles to each other. This will normally occur, except in conditions where diffuse double layers develop. Forces between domains will also involve electrical interactions, but these may be weaker because of fewer contacts and greater water film thickness between them. However, they will often be supplemented by natural organic polymers, forming interparticle bonds, or by precipitates of inorganic hydroxides which act as interparticle cements. Between microaggregates it is probable that occasional bonding of this type, acting at relatively few contact points between them, is the only source of stability. Thus aggregate stability can usually be rather closely related to content of organic matter, except in those soils where adequate cementation between domains and microaggregates is created by hydrous oxides.

At one time red tropical soils were all considered to consist of aggregates strongly cemented by iron oxides. It is now known that this is not true, and many tropical red earths contain only structurally inactive iron oxides, at least in the surface soil (Deshpande *et al.*, 1968). The weakly acid Alfisols and Ultisols of the tropics appear frequently to contain much of their iron oxide as small hydrohaematite granules. These soils are also often low (<2 per cent organic carbon) in organic matter, and consequently of low aggregate stability. The highly acid Oxisols normally contain active aluminium oxides as cementing agents, and these structures may be more stable.

Many tropical soils developed on highly basic rocks also appear to contain structurally active oxides, probably amorphous iron and aluminium hydroxides, and these soils, as well as Andosols containing allophane developed in volcanic ash, are of exceptional structural stability, although, as shown in Chapter 3.3, their low cohesion may make them highly erodible. The soils developed on basic rocks and volcanic ash also have very high organic matter contents for tropical soils, and their aggregate stability will in part also be due to this factor.

Whether or not aggregates and microaggregates are cemented by hydrous oxides, there is usually sufficient aluminium amongst the surface ions in the more acid soils to ensure flocculation of the clay. Consequently, translocation of clay into the subsoil and reduction of subsoil permeability in Oxisols occurs slowly if at all, and subsoil permeabilities tend to be high. On the less acid soils translocation does occur, and the recognition of the current occurrence of this phenomenon in many 'tropical red earths' is important in relation to their classification, particularly if the classification is to reflect their liability to erosion.

2.2.5. MEASUREMENT OF 'STRUCTURE' AND STRUCTURAL STABILITY

Many methods for describing the structure of soils have been devised, but no adequate methods for its measurement. Pore-size distribution, continuity and tortuosity are the most appropriate measurements to make, but usually these are dependent on measurements of gas or water flow in the soil. Thus it is usually preferable to obtain flow measurements—infiltration rates, or hydraulic conductivities—and use these as a measure of structure.

Special problems are posed by swelling soils, or non-swelling soils where structure is readily changed, when different cultural practices are followed. Changes in flow characteristics due to specific stresses are perhaps the most significant measurement of structural stability. However, because of the importance of surface soils, and the 'optimum' condition associated with the presence of a large proportion of the particles in aggregates of 2–5 mm size, much effort has been devoted to the arbitrary measurement of 'aggregate stability' of samples of this size. Different stresses are applied, either mechanically, by wet or dry sieving, by immersion of dry aggregates in water, or by increasing the inherent swelling pressures by sodium saturation. Each of these can give an arbitrary measure of relative stabilities, which may be related to differences in field behaviour (Williams *et al.*, 1966). However, in relation to erodibility, the stresses associated with raindrop impact are of most significance, and stability to stresses applied by rainfall simulation are probably the most useful criteria (e.g. Bruce-Okine and Lal, 1975).

2.2.6 ENHANCED EROSION HAZARD ASSOCIATED WITH CULTIVATION

Removal of the natural vegetative cover of a tropical soil exposes it to direct raindrop impact. The impact stress of these raindrops may be as much as 10^5 Nm^{-2}, which is probably in excess of the strength of most aggregates when wet. Trampling and cultivation implements exert similar or larger forces. Consequently, rapid breakdown of surface structure is to be expected, even when natural vegetation is newly cleared. Only a few, if any, soils will be able to withstand the compactive forces to which the soils are exposed. Furthermore, the absence of factors able to recreate an aggregated structure, as for instance freezing and thawing does with many temperate soils, means that the deterioration is more or less permanent, unless the soil is put under fallow vegetation for an extended period, when the soil fauna redevelop and recreate the structure. Tillage may recreate the structure, but this will not establish any stability, which has to be provided by organic matter put into place by the fauna.

The disaggregation of the surface soil leads to much reduced water entry, and the newly isolated soil particles are readily detached and removed in runoff water.

This deterioration in soil condition is inevitable unless the soil can be protected from the disruptive stresses, and the fauna which create the favourable structure maintained. Hand-clearing, zero tillage, mulching and mixed cropping designed to keep a crop cover above the soil allow this to be done. Mechanized farming involving burial of crop residues will always tend to increase the erosion hazard. It may be possible to control the erosion by appropriate engineering structures, but the skill and care required for their construction and maintenance are often lacking, and in food crop production their cost is likely to be uneconomic.

2.2.7. CONCLUSIONS

Under natural vegetation most tropical surface soils have very porous, permeable structures. The stability of this structure is often inadequate to resist the large stresses to which they are subject on exposure for cultivation. Although hydrous oxides of iron and aluminium are often present in these soils, the iron oxides at least are usually not structurally active. Thus permeability is lost on exposure, and runoff and erosion occur readily, unless methods such as zero tillage are used to preserve the existing structure, and mulching and intercropping to maintain it.

REFERENCES

Bruce-Okine, E., and R. Lal, 1975, Erodibility as determined by raindrop technique, *Soil Science*, **119**, 149–157.

Cohron, G. T., 1971, Forces causing soil compaction, in *Compaction of Agricultural Soils*, Amer. Soc. Agric. Eng., St. Joseph, Michigan.

COPR, 1974 and 1975, Interim reports, Pesticide Residues Research Project, IITA, COPR, London.

Deshpande, T. L., D. J. Greenland, and J. P. Quirk, 1968, Changes in soil properties associated with the removal of iron and aluminium oxides, *J. Soil Sci.*, **19**, 108–122.

Greenland, D. J., 1971, Changes in the nitrogen status and physical condition of soils under pasture, *Soils & Fertilisers*, **34**, 237–251.

Lal, R., 1975, Role of mulching techniques in tropical soil and water management, IITA Tech. Bull. No. 1.

Lal, R., and Mensah Bonsu, 1975, Field and laboratory determinations of hydraulic conductivity of an alfisol in Nigeria, *Proc. Latin American Congress of Soil Science, Medellion, Colombia.*

Marshall, T. J., 1962, Soil structure, *Proc. Comm. IV and V, Int. Soc. Soil Sci., New Zealand*, **1**, 243–257.

Sanchez, P. A., 1974, Report on soil and agronomy studies at Caramaguas, Peru, Univ. of Wisconsin.

Trouse, A. C., and L. D. Baver, 1965, Data quoted in L. D. Baver, W. L. Gardner, and W. R. Gardner, *Soil Physics*, 4th edn, Wiley, New York.

van der Weert, 1974, Infulence of mechanical clearing on soil conditions and the resulting effects on root growth, *Trop. Agric.* (*Trin.*), **51**, 325–333.

Wilkinson, G. E., and P. O. Aina, 1976, Infiltration of water into two Nigerian soils under secondary forest and subsequent arable cropping, *Geoderma*, **15**, 51–59.

Williams, B. G., D. J. Greenland, G. R. Lindstrom, and J. P. Quirk, 1966, Techniques for the determination of the stability of soil aggregates, *Soil Sci.*, **101**, 157–163.

2.3

Water Entry and Movement in Relation to Erosion

W. R. GARDNER

Surface water flow is the culprit in soil erosion, but it is subsurface flow which, along with the precipitation, governs the surface flow process. The capacity of the soil mantle to absorb water is finite, and whenever the rainfall rate exceeds the infiltration capacity the result is surface runoff. When the rainfall rate is less than the infiltration capacity, gravity and the affinity of soil particles for water molecules combine to direct the impinging water droplets into the subsurface flow system as rapidly as they strike the surface. It is in the nature of the infiltration process that the rate of precipitation or the infiltration capacity of the soil—whichever is the lesser—determines the infiltration rate. Anytime that the soil limits the infiltration rate, the excess precipitation, beyond that amount which can be detained in surface depressions, runs downhill over the soil surface, usually acquiring increasing potential for soil erosion as it transforms from sheet flow to rill flow.

Neither time nor space permit an adequate discussion of infiltration and surface runoff, and this chapter is really in the nature of an annotated bibliography. Beyond the effect of surface stability or erodibility and surface topography, the thickness of the surface water film influences erosion as it cushions the raindrop impact and entrains soil particles dislodged from the soil surface (Mutchler and Larson, 1971; and Meyer, 1971).

The surface flow system is two-dimensional, while the infiltration and subsurface flow in the unsaturated soil profile is usually treated as one-dimensional vertical flow. Further down in the saturated groundwater system two or all three dimensions must be considered. The very important problem of linking the one-dimensional system to the multidimensional geometry above and below it, is only now receiving much attention from soil physicists. It is an extremely difficult problem due to the heterogeneous nature of soils. We know that certain areas produce more runoff than others. These areas, however, may not coincide with the most erosive soil surfaces or may not be uphill from erosive spots. Thus, we must keep in mind that our theories at most can be useful guides and can be of great value in helping us decide what measurements to make and where and how to interpret those measurements. At some stage we must go to the field, and the greatest challenge facing soil physicists today is to scale up our theories and our measurements from the dimensions of laboratory soil columns to the dimensions of a field or a watershed.

2.3.1 THE PHYSICS OF INFILTRATION

Unless it be soil drainage, no soil hydrological process has received the attention that has been devoted to the infiltration process. The first significant applications of the equation describing flow in unsaturated soils (Klute, 1952; Philip, 1957 and subsequently) were to the infiltration process. A review of early mathematical literature on infiltration has been written by the author (Gardner, 1967) and a comprehensive review of the mathematical treatment of the process is given by Philip (1969). Parr and Bertrand (1959) wrote a comprehensive and useful review of the early English language literature. (This USDA report did not receive wide distribution. It does not merit reproduction now, but the present author would be willing to send copies of the bibliography to anyone requesting it.) This chapter will not attempt to review this voluminous subject, but will consider in a semi-quantitative way the basic elements which are important to an understanding of the role which various factors play in infiltration. Mathematical details of the solution to the flow equation (a form of the Fokker–Planck equation) are given in the 1969 paper by Philip and reviews by Swartzendruber (1969) and Swartzendruber and Hillel (1973).

2.3.1.1. Infiltration Equations

Water filters downwards into the soil in response to two driving forces—gravity and the force of attraction between water molecules and soil particle surfaces. It is generally observed that the infiltration capacity of the soil is finite and tends to decrease during the course of a rainstorm and recover during periods of no rainfall. A number of empirical formulae describing the time-course of infiltration have been used. The best-known of these are probably that of Horton (1933):

$$i = i_\infty + (i_0 - i_\infty)\exp(-bt) \tag{2.1}$$

and Kostiakov (1932):

$$I = \int i\,dt = Kt^a \tag{2.2}$$

where i is the infiltration rate at any time t; i_0 is the initial infiltration rate at $t = 0$; and i_∞ is the final infiltration rate at infinite time. The constants b, K, and a are parameters which depend upon the physical properties of the soil profile. The constants in equations (2.1) and (2.2) do not lend themselves readily to precise physical interpretation and must always be measured in the field. It is assumed that water is always supplied at a rate sufficient to maintain a thin film of water at the soil surface. Equation (2.1) allows us to separate two aspects of the flow process. Initially, if the soil profile is not completely saturated with water, the capillary forces add to the gravitation force to determine the driving force for water movement. When this force is multiplied by the hydraulic conductivity one obtains the initial infiltration rate i_0. As time goes by the wetting front penetrates since the capillary forces become less and less important until, eventually, a steady infiltration rate due solely to the gravitation gradient is achieved, the final rate i_∞. Other factors than the decreasing matric potential gradient may contribute to the decreasing infiltration rate and some of these will be discussed in the next section.

For a rigorous treatment of the infiltration process one starts with the flow equation and, assuming a homogeneous soil profile initially at a uniform water content, one can derive an equation for infiltration in terms of an infinite series in powers of $t^{1/2}$ (Philip, 1957):

$$I = St^{1/2} + At + Bt^{3/2} + \cdots \tag{2.3}$$

where S, A, and B are coefficients which depend upon the soil properties as well as the initial water content. Of particular interest is the parameter S, which Philip has called the sorptivity. It is proportional to the increase in soil water content during infiltration and to the square root of the soil water diffusivity, when properly averaged. Equation (2.3) does not hold for large values of t, since then the neglected terms become too large. A convenient approximation is achieved by replacing all terms beyond the first by Kt, i.e.:

$$I = St^{1/2} + Kt \tag{2.4}$$

where K is the hydraulic conductivity of the soil. Equation (2.4) now has the correct form at both small and large values of t and is not seriously in error at intermediate times.

2.3.2. COMPLICATING FACTORS

The empirical equations (2.1) and (2.2) can describe the infiltration to an adequate degree of approximation but give us little clue as to how soil properties and processes might influence infiltration. The flow equation approach can deal with any complication which we may wish to include, but the necessity to use numerical methods (and high-speed digital computers) prevents our ready understanding, unless each case is solved separately and then, unless we have been extremely clever in our approximations, we pay for and receive much more detail than we need.

2.3.2.1. The Green and Ampt Equation

A very simple model of infiltration which allows both physical interpretation and numerical approximation is one of the earliest, which was developed by two

Australians (Green and Ampt, 1911). In their approach one assumes that the pores of a soil can be thought of as a bundle of capillaries in which water advances in response to the capillary force (as well as gravity) with a very well-defined wetting front. The soil is assumed to have a constant hydraulic conductivity behind the wetting front and it is further assumed that the soil suction or soil moisture tension at the wetting front can be represented by a single value, characteristic of the soil pore size. With these modifications to Darcy's law, one obtains:

$$i = dI/dt = K\frac{(H_0 + H_f \cos\alpha + L - P)}{L} \tag{2.5}$$

where i is the instantaneous infiltration rate; I is the total infiltration up to time t; H_0 is the depth of water ponded on the soil surface; H_f is the effective suction at the wetting front; α is the wetting angle, usually zero; L is the depth to the wetting front; and P is the air pressure (gauge) at the wetting front. It follows from our assumptions that

$$I = \Delta\theta L \tag{2.6}$$

where $\Delta\theta$ is the difference between the initial and final water contents, on a volumetric basis.

The effective suction at the soil surface is generally less than a few tens of centimetres, and is less for sandy soils than finer textures.

2.3.2.2. Other Factors

We have included in equation (2.5) several factors which were not in Green and Ampt's equation. If the wetting angle is not zero, as is the case for hydrophobic soils, $\cos\alpha$ is less than one and may be negative, in which case a positive pressure is required to force water into the soil. It is usually assumed that air moves readily out of the soil pores ahead of the infiltrating water. This requires a pressure gradient which may necessitate inclusion in equation (2.5) of a pressure, P. This problem is discussed by Morel-Seytoux and co-workers (1972; 1974), Van Phuc and Morel-Seytoux (1972), Vachaud *et al.* (1973), Dixon and Linden (1972) and Linden and Dixon (1973), as well as Garner *et al.* (1969) and Peck (1965).

A very common physical property of a soil profile is a thin surface layer or crust with lower permeability than the soil horizons below the surface. Equation (2.5) requires modification in order to treat this case (Hillel and Gardner, 1970a) and the degree of approximation is less satisfying. If one knows the crust resistance and the soil unsaturated permeability as a function of suction, the steady-state problem is relatively straightforward (Hillel and Gardner, 1969; Kutilek, 1974). One consequence of a crust is the seemingly paradoxical result that the flow through a coarse soil is less than that for a finer-textured soil, all other factors being equal. This is caused by the hydraulic head loss across the crust, which results in a suction at the crust–soil interface. This suction desaturates the underlying soil. At the same suction a sand with a greater volume of coarse pores is less permeable than a clay, since the large pores are the first to empty.

Artificial surface crusts can be used to good effect to determine the soil unsaturated conductivity (Hillel and Gardner, 1970b; Bouma *et al.*, 1971). If the crust resistance changes during infiltration, as is often true of unstable soils, mathematical treatment of the problem is very difficult.

Up to this point, we have considered only that case where the rainfall rate exceeds the soil infiltration rate at all times. In the early minutes of a rainstorm, the soil infiltration capacity may be very high and may exceed the precipitation rate. In this instance we would like to know how long a given rainfall rate can be accommodated by the soil before runoff occurs. Mein and Larson (1973) and Smith (1973) have modified equation (2.5) to estimate this. A more rigorous analysis has been worked out by Rubin and colleagues (Rubin and Steinhardt, 1964; Rubin *et al.*, 1964). For erratic rainfall, with brief periods of little or no precipitation interspersed with rates higher than the soil infiltration capacity, the problem is even more complicated. Reeves and Miller (1975) have devised a very simple approach to this problem. They also include hysteresis in their treatment.

2.3.3. A FINAL WORD

The value of a theoretical analysis lies not so much in prediction of actual runoff but in identification of the major physical factors involved. Field measurements are always necessary to validate the theory. They are not usually easy to make, nor totally satisfying (Jeppson *et al.*, 1975; Talsma, 1974a, b; Talsma and Parlange, 1972). The familiar infiltration ring measurements are not truly one-dimensional (Swartzendruber and Olson, 1961). Much remains to be done to relate the structural characteristics of soil profiles to runoff and erosion. Some ambitious field studies coupled with continuing mathematical analysis of the system are required.

REFERENCES

Bouma, J., D. I. Hillel, F. D. Hole, and C. R. Amerman, 1971, Field tests of unsaturated hydraulic conductivity by infiltration through artificial crusts, *Soil Sci. Soc. Am. Proc.*, **35**, 362–364.

Brustkern, R. L., and H. J. Morel-Seytoux, 1970, Analytical treatment of two phase infiltration, *J. Hyd. Div. Am. Soc. Civil Engr.*, **96** (HY12), 2535–2548.

DeBano, L. F., 1971, The effect of hydrophobic substances on water movement in soil during infiltration, *Soil Sci. Soc. Am. Proc.*, **35**, 340–343.

Dixon, R. M., and D. R. Linden, 1972, Soil air pressure and water infiltration under border irrigation, *Soil Sci. Soc. Am. Proc.*, **36**, 948–953.

Dixon, R. M., and A. E. Peterson, 1971, Water infiltration control: a channel system concept, *Soil Sci. Soc. Am. Proc.*, **35**, 868–873.

Epstein, E., and W. J. Grant, 1967, Soil losses and crust formation as related to some soil physical properties, *Soil Sci. Soc. Am. Proc.*, **31**, 547–550.

Gardner, W. R., 1967, Development of modern infiltration theory and application in hydrology, *Trans. Am. Soc. Agr. Engr.*, **10**, 379–381, 390.

Garner, D. E., J. K. Donaldson, and G. S. Taylor, 1969, Entrapped soil air in a field site, *Soil Sci. Soc. Am. Proc.*, **33**, 634–635.

Green, R. E., R. J. Hanks and W. E. Larson, 1964, Estimates of field infiltration from numerical solution of the moisture flow equation, *Soil Sci. Soc. Am. Proc.*, **28**, 15–19.

Green, W. H., and G. A. Ampt, 1911, Studies on soil physics. I. The flow of air and water through soils, *J. agric. Sci.*, **IV**, 1–24.

Gumbs, F. A., and B. P. Warkentin, 1972, The effect of bulk density and initial water content on infiltration in clay soil samples, *Soil Sci. Soc. Am. Proc.*, **36**, 720–724.

Gupta, R. P., and D. Swartzendruber, 1964, Entrapped air content and hydraulic conductivity of quartz sand during prolonged liquid flow, *Soil Sci. Soc. Amer. Proc.*, **28**, 9–15.

Hillel, D., and W. R. Gardner, 1969, Steady infiltration into crust-topped profiles, *Soil Sci.*, **108**, 137–142.

Hillel, D., and W. R. Gardner, 1970a, Transient infiltration into crust-topped profiles, *Soil Sci.*, **109**, 69–76.

Hillel, D., and W. R. Gardner, 1970b, Measurement of unsaturated conductivity and diffusivity by infiltration through an impeding layer, *Soil Sci.*, **109**, 149–153.

Horton, R. E., 1933, The role of infiltration in the hydrologic cycle, *Amer. Geophys. Union*, **14**, 446–460.

Jeppson, R. W., W. J. Rawls, W. R. Hamon, and D. L. Schreiber, 1975, Use of axysymmetric infiltration model and field data to determine hydraulic properties of soils, *Water Resources Res.*, **11**, 127–138.

Kijne, J. W., 1967, Influence of soil conditioners on infiltration and water movement in soils, *Soil Sci. Soc. Am. Proc.*, **31**, 8–13.

Klute, A., 1952, A numerical method for solving the flow equation for water in unsaturated materials, *Soil Sci.*, **73**, 105–116.

Kostiakov, A. N., 1932, On the dynamics of the coefficient of water percolation in soils and the necessity for studying it from a dynamic point of view for purposes of amelioration, *Trans. 6th Comm. Inter. Soil Sci. Soc.*, Russian Part A, 17–21.

Kutilek, M., 1974, Infiltration into crust-topped soil, *Polish J. Soil Sci.*, **7**, 3–7.

Linden, D. R., and R. M. Dixon, 1973, Infiltration and water table effects of soil air pressure under border irrigation, *Soil Sci. Soc. Am. Proc.*, **37**, 94–98.

McIntyre, D. S., 1958a, Permeability measurements of soil crusts formed by raindrop impact, *Soil Sci.*, **85**, 185–189.

McIntyre, D. S., 1958b, Soil splash and the formation of surface crusts by raindrop impact, *Soil Sci.*, **85**, 261–266.

Mein, R. G., and C. L. Larson, 1973, Modeling infiltration during a steady rain, *Water Resources Res.*, **9**, 384–394.

Meyer, L. D., 1971, Soil erosion on upland areas, Chapter 27, In H. W. Shen (Ed.), *River Mechanics*, Vol. 2, H. W. Shen Publ., Fort Collins, Colorado.

Morel-Seytoux, H. J., 1972, Two-phase flows in porous media, *Adv. Hydroscience*, **9**, 119–202.

Morel-Seytoux, H. J., and J. Khani, 1974, Derivation of an equation of infiltration, *Water Resources Res.*, **10**, 795–800.

Mutchler, C. K., 1971, Splash droplet production by waterdrop impact, *Water Resources Res.*, **7**, 1024–1030.

Mutchler, C. K., and C. L. Larson, 1971, Splash amounts from waterdrop impact on a smooth surface, *Water Resources Res.*, **7**, 195–200.

Parr, J. F., and A. R. Bertrand, 1959, A literature review of water infiltration into soils, USDA, ARS, SWCRD Special Report No. 83.

Peck, A. J., 1965, Moisture profile development and air compression during water uptake by bounded porous bodies. 3. Vertical columns, *Soil Sci.*, **100**, 44–51.

Philip, J. R., 1954, An infiltration equation with physical significance, *Soil Sci.*, **77**, 153–157.

Philip, J. R., 1957, The theory of infiltration. 1. The infiltration equation and its solution, *Soil Sci.*, **83**, 345–357.

Philip, J. R., 1968, Absorption and infiltration in two- and three-dimensional systems, *UNESCO Symposium on Water in the Unsaturated Zone*, Vol. I, pp. 440–450.

Philip, J. R., 1969, Theory of infiltration, *Adv. Hydroscience*, **5**, 215–299.

Reeves, M., and E. E. Miller, 1975, Estimating infiltration for erratic rainfall, *Water Resources Res.*, **11**, 102–110.

Rubin, J., and R. Steinhardt, 1964, Soil water relations during rain infiltration. III. Water uptake at incipient ponding, *Soil Sci. Soc. Am. Proc.*, **28**, 614–618.

Rubin, J., R. Steinhardt, and P. Reiniger, 1964, Soil water relations during rain infiltration. II. Moisture content profiles during rain of low intensities, *Soil Sci. Soc. Am. Proc.*, **28**, 1–5.

Smith, R. E., 1973, Comments on 'Modeling infiltration during a steady rain' by R. G. Mein and C. L. Larson, *Water Resources Res.*, **9**, 1475–1477.

Swartzendruber, D., 1969, The flow of water in unsaturated soils, In R. J. M. deWiest (Ed.), *Flow through Porous Media*, pp. 215–292.

Swartzendruber, D., and D. Hillel, 1973, The physics of infiltration, In A. Hadas, D. Swartzendruber, P. E. Rijtema, M. Fuchs, and B. Yaron (Eds.), *Physical Aspects of Soil Water*, Springer-Verlag, New York, Heidelberg, Berlin, pp. 3–15.

Swartzendruber, D., and T. C. Olson, 1961, Sand-model study of buffer effects in the double-ring infiltrometer, *Soil Sci. Soc. Am. Proc.*, **25**, 5–8.

Tackett, J. L., and R. W. Pearson, 1965, Some characteristics of soil crusts formed by simulated rainfall, *Soil Sci.*, **99**, 407–413.

Talsma, T., 1974a, Infiltration in field soils, *Trans 10th Int. Congr. Soil Sci.* (*Moscow, 1974*), **I**, 68–74.

Talsma, T., 1974b, The effect of initial moisture content on infiltration quantity and redistribution of water, *Aust. J. Soil Res.*, **12**, 15–26.

Talsma, T., and J.-Y. Parlange, 1972, One-dimensional vertical infiltration, *Aust. J. Soil Res.*, **10**, 145–150.

Vachaud, G., M. Vauclin, and M. Wakil, 1973, Effects of air pressure on water flow in an unsaturated stratified vertical column of soil, *Water Resources Res.*, **9**, 161–173.

Van Phuc, Le, and H. J. Morel-Seytoux, 1972, Effects of soil air movement and compressibility on infiltration rates, *Soil Sci. Soc. Am. Proc.*, **36**, 237–241.

2.4

Cultural Practice and Changes in Catchment Hydrology: A Review of Hydrological Research Techniques as Aids to Development Planning in the Humid Tropics

K. A. EDWARDS

2.4.1. INTRODUCTION

The humid tropics are regions of heavy rainfall of which a large proportion falls in storms of high intensity. Although the potential for soil erosion is high, the indigenous tropical rain forest once typical of this zone produced soils with high infiltration characteristics which were protected by the closed canopy from the mechanical action of the intense rainfall.

Population pressure has gradually forced the progressive clearing of the indigenous forest for agriculture. The development of plantation agriculture has had a similar result. The protection now offered to the soil against high insolation and heavy rainfall varies greatly with the type of crop and the system of cultivation. While the artificial loosening of the soil by tillage leads to temporary maintenance of high infiltration rates, unless the soil is stabilized, soil erosion leads to a decline in fertility, a deterioration in the quality of water supplies and a change in the hydrological response of a river to rainfall.

The early stages of the accelerating progress of erosion are not always easily identifiable, especially on the less steep slopes, and the problems of the quantitative assessment of the severity of the erosion hazard resulting from changes in cultural practices and the evaluation and comparison of remedial treatments are ones to which hydrological techniques of catchment research can be applied. At the same time, modifications to the total volume and seasonal distribution of streamflow can also be analysed together with any tendency for the flood peaks to increase.

The catchment is a natural unit within which the effects of a number of physical parameters (slope, aspect, altitude and elevation) can be integrated to give quantitative estimates of changes in the water balance or in sediment yield which can be immediately applied to practical land-use questions. Plot experiments, which are particularly favoured for the study of soil erosion, suffer severely from sampling problems and edge effects when used for water balance studies, and extrapolating to a larger scale is frequently impossible (Boughton, 1967; Martinelli, 1964). Although plot experiments may give more insight into the nature of physical processes, any physical models so derived must ultimately be verified and assessed in relation to catchment data (Ward, 1971). There is a strong case, therefore, for integrating these two approaches within carefully designed catchment experiments aimed at investigating the physical nature of the erosion process and measuring the hydrological effects of land-use change.

This chapter reviews some of the catchment studies relevant to the development of soil conservation and management guidelines in the humid tropics. It deals with design criteria for small experimental catchments and the special problems associated with this type of hydrological research.

2.4.2. EXISTING CATCHMENT STUDIES

Quantitative studies of the effects of cultural change on the hydrological cycle which refer specifically to the humid tropics are rare. Many of the accounts of temperate latitude research, however, discuss at length the problems of catchment selection, network design, instrumentation, data collection and data processing. Although it is not recommended that experimental results are extrapolated from one climatic region to another, a survey of existing experiments will give valuable insight into the improvements in research techniques and equipment which have taken place over the last twenty years.

Historically, catchment research has been a recent innovation outside the USA apart from some notable exceptions (Engler, 1919; Hirata, 1929). Many of the early experiments, however, were simply measurements of rainfall and runoff, with no attempt being made to interpret the measured changes in terms of the physical processes of the hydrological cycle. A major change occurred with the development of empirical evaporation formulae (Thornthwaite, 1948; Penman, 1948) which could be used to calculate seasonal water balances. Increasing attention to the problems of conserving water resources stimulated the development of methods of measurement and instruments for hydrological variables. The International Hydrological Decade 1965–1975 led to an interchange of information and results which has emphasized the problems of representativeness of catchments and the extrapolation of results to other regions. Da Costa and Jacquet (1965) at the start of the IHD listed 726 representative and experimental basins. Ibbitt (1974) estimates that there are now over 3000 in the world. Certain climatic zones are not well served, however, and there is a need in particular to establish representative and experimental basins in those regions such as the humid tropics where land-use change presents pressing soil and water conservation problems.

Of the longer established studies, those of ORSTOM (Rodier and Auvray, 1965; Rodier and Vuillaume, 1972) in West Africa and EAAFRO (Pereira *et al.*, 1962; Dagg and Blackie, 1965; Edwards and Blackie, 1975) are perhaps the most relevant, although many of the former are concerned with drier zones and the latter refer mainly to the montane humid forest.

Results from other regions may be found in the proceedings of the Budapest symposium (IAHS, 1965a) and the Wellington symposium (UNESCO/IAHS, 1972). A third international symposium dealing with results from experimental and representative basins took place in Tokyo in December 1975. Other International Association of Hydrological Sciences publications particularly relevant to catchment research are the proceedings of the symposia of Hannoversh-Munden (IAHS, 1959), Bari (IAHS, 1962), Quebec (IAHS, 1965b), and Reading (IAHS, 1970).

UNESCO have also a series of publications entitled 'Studies and Reports in Hydrology' of which several refer specifically to catchment research. For example, Toebes and Ouryvaev (undated) is a guide to research and practice relating to representative and experimental basins. There are also the proceedings of the Warsaw symposium on mathematical modelling, which are particularly useful (UNESCO, 1971).

Research watersheds were also the subject of a symposium in Ontario (University of Guelph, 1965), and in Pennsylvania there was an international conference on forest hydrology from which much useful guidance can be obtained (Sopper and Lull, 1967). Penman (1963) has summarized the results of some of the older and better-known catchment experiments, while Ward (1971) gives a comprehensive bibliography and review of small watershed (catchment) experiments.

Other review papers include Leopold (1972) and Hibbert (1967), who discusses the results of 39 studies of the effects of forest excision. The American Geophysical Union published an inventory of representative and experimental watershed studies conducted in the United States (1965), and recommendations have been made to update the world inventory of Da Costa and Jacquet (1965) as part of the continuing International Hydrological Programme (UNESCO/WMO, 1974).

Not all the review papers support the proliferation of catchment experiments. The approach has come under severe criticism from Ackermann (1966) and Reynolds and Leyton (1967) amongst others, although most of their arguments have been countered by Hewlett *et al.* (1969). Catchment experiments are costly and do not always produce results which can be readily applied to areas other than the catchments themselves. Many of the experiments were ill-conceived and mismanaged according to Ackermann (1966), and their results are often contradictory. To illustrate the difficulty of extrapolating results from one environment to another, nearly all the US forest-cutting experiments show increases in total water yield following clear-felling (Liebermann and Hoover, 1951; Hewlett and Hibbert, 1961; Reinhart and Eschner, 1962; Lull and Reinhart, 1967). On the other hand, Rakhmanov (1966), considering data from the USSR, concludes that forests increase precipitation, decrease evaporation and tend to increase the water resources of an area.

While the above references will give valuable illustrations of research techniques and methods of analysis, there is a danger in attempting to use results out of context in areas where significant differences in the hydrological cycle can change the water balance. For this reason, where such studies can contribute towards water resources and land-use planning, there is merit in designing carefully controlled catchment experiments both as basic data sources and as investigations into the hydrological effects of land-use change.

2.4.3. THE DESIGN OF CATCHMENT EXPERIMENTS

The first priority in the design of catchment experiments is the definition of research objectives. In terms of establishing guidelines for agricultural development in the humid tropics, those experiments should be aimed at:

(a) measuring sediment yield and the severity of the erosion hazard;
(b) assessing the effects of cultural change on total water yield, on its seasonal distribution and on storm runoff; and
(c) assessing the efficiency of remedial measures in combating erosion.

In (b) above, the size of the catchment is important. Penman (1966) has demonstrated a method for determining the minimum size based on 'edge effects' if the catchment is to yield meaningful water-use data. The maximum size, on the other hand, is often predetermined by the limited availability of suitable watertight catchments and the manpower requirements for operating the instrument networks. The relationship between catchment size and peak runoff is complex and not fully understood (Wolf, 1966). If deforestation is suspected of increasing flood peaks in the region under study, it is important to choose catchments similar in size and shape for direct comparisons to be made.

Since it is impossible to control the input to the hydrological cycle, an experiment must be of sufficient duration adequately to sample the range of storm sizes and rainfall intensities required. In addition, a calibration period is needed to check whether the catchments are watertight and to establish the relationship between the experimental catchments and the control.

The preferred experimental method is to select two catchments, similar in size, aspect, altitude, soil type and vegetation cover, one of which is destined to remain as a control while the other undergoes a change in land use. Both catchments are instrumented to measure the hydrological variables to a precision in keeping with the expected magnitude of the effects of the change in land use. After a calibration period of at least two water-years, the change is effected and the variables are monitored until either a stable state is reached or a sufficient indication of the rate of change of the hydrological response is obtained.

2.4.4. NETWORK DESIGN

A fundamental problem of catchment research is to obtain mean areal values of hydrological variables from point measurements. Streamflow is concentrated at a single outfall, where a representative measurement can easily be made if no water bypasses the measuring section. Evaporation, rainfall and soil moisture, however, are spatially variable and a sampling method is required to estimate the areal mean.

In the case of small catchments (<25 km^2), where the solar radiation input does not vary due to persistent cloud in part of the catchment, a single meteorological station, situated where there is sufficient fetch over the vegetation type for equilibrium conditions to be established, will suffice for potential evaporation measurements. Actual evaporation varies with vegetation height, albedo, aspect and soil moisture content, but unless it is intended to develop models with spatially distributed inputs, the problems associated with using actual evaporation as a catchment variable outweigh any advantages compared with using potential evaporation as a 'lumped' input for small catchment models.

Rainfall is a most difficult variable to measure accurately, especially in forested catchments. Point measurements are subject to significant errors due to the effect of exposure above the ground on raingauge catch (Poncelet, 1959; Rodda, 1967). Fortunately, in tropical regions the conditions of high wind speed combined with small drop size are rare, and errors due to this effect are small.

Rainfall in a forest should be measured by gauges raised to canopy height by means of towers or platforms on trees (Pereira *et al.*, 1962). Such gauges are expensive to install and difficult to maintain and, if conditions permit, standard gauges placed in clearings give more consistent results. The minimum size of such clearings has been investigated by McCulloch (in Pereira *et al.*, 1962), who found that clearings equivalent to 45° screening (i.e. with a radius equal to the height of the trees) were adequate in the montane forests of East Africa. Salter (1919) and the Pacific Southwest Inter-Agency Committee (1962) recommend that the angle of elevation of the top of the nearest trees should not exceed 30°, thus allowing for further growth.

Estimates should also be made of the interception storage of the forest canopy in order to calculate the total evaporative loss, i.e. transpiration and evaporation of intercepted water. Throughfall and stemflow must be estimated, therefore, for the whole catchment, and normally this poses difficult sampling problems (Reynolds and Leyton, 1963). To avoid an excessively large number of throughfall gauges troughs are often used, although the increased quantities of water call for special measuring devices (Rutter, 1963). A good example of a complete interception experiment will be found in Hamilton and Rowe (1949), together with a useful but early bibliography.

The determination of the optimum density for raingauge networks has been the subject of a number of studies (Rycroft, 1949; McCulloch, 1961; Panchang and Narayan, 1962; Herbst and Shaw, 1969). All the statistical techniques require data from existing networks to obtain an estimate of the variance of point rainfall, and in forested catchments the installation of a dense network initially and the subsequent reduction of the number of gauges can be costly. At present, however, where no representative data are available for analysis, there is no alternative if valid estimates of the precision of the areal mean are required (Huff and McNeil, 1957; McGuiness, 1963; Nicks, 1965).

Clarke and Edwards (1973) have shown that many of the statistical methods used in network design can be reduced to the application of the technique of analysis of variance to rainfall data. In using the residual variance from the analysis of variance of existing data to determine raingauge densities, however, there is no guarantee that the mean areal rainfall for any future period will be estimated with the same precision unless the interaction between gauges and periods is taken into account.

Even if the optimum number of gauges can be decided upon, the problem of positioning the gauges in a sampling network remains. Random sampling will give valid estimates of the variance but not necessarily the best estimate of the mean. Systematic sampling, with its comprehensive spatial coverage, usually gives a less biased estimate of the mean, and a combination of the two techniques, such as stratified random sampling based on altitude or proximity (McCulloch, 1961), is often favoured.

In such a network, computing the area mean is simplified if the number of gauges in each stratum is proportional to the area of the stratum. The areal mean is then the arithmetic mean of all gauges, and if the density of the raingauge network is determined by the above techniques, no increase in accuracy will result from the application of more sophisticated techniques of estimating the areal mean.

Problems arise when gauges are out of commission for some length of time. In this case, Thiessen polygons, polynomial rainfall surfaces (Mandeville and Rodda, 1970; Edwards, 1972), multiquadric techniques (Lee *et al.*, 1974) and time series analysis (Hutchinson and Walley, 1972) will improve the estimates of mean areal rainfall with irregularly spaced data. In this context, however, it is important to consider that all linear models can be reduced to a set of raingauge weights (Chidley and Keys, 1970) and that these will differ insignificantly from unity in a network designed on a self-weighting stratified random sampling basis of adequate density when all gauges are working. Furthermore, allowance can easily be made for missing data by estimating stratum means separately from the reduced number of gauges and, if these are still statistically valid, no benefit will be obtained from the alternative techniques.

Soil moisture is also highly variable in space depending on the distribution of soil types, slope, aspect, position on the catena and vegetation cover. With the advances in neutron scattering techniques (Bell, 1969), point measurement of soil moisture content in a given profile is rapid but network densities are still limited by the number of sites an observer can visit in a day.

The preceding remarks on rainfall sampling with respect to the variance of point measurements are equally applicable to soil moisture, except that the latter is usually more spatially variable than rainfall. This calls for denser networks of neutron access tubes, and a compromise must inevitably be sought between desirable precision, availability of suitable sites and the cost of installation of access tubes. In highly variable soils, the effort involved in calibrating individual soil profiles must also be taken into account.

2.4.5. SPECIAL PROBLEMS

2.4.5.1. Groundwater

In permeable catchments, the assessment of changes in groundwater storage is essential to an understanding of the seasonal water balance. Normally, direct measurements of the static groundwater level can only be made from boreholes and wells, which are much too costly to be installed purely for catchment research. It is possible, however, to estimate the level of groundwater storage from the recession curve of the hydrograph (Blackie, 1972a). Alternatively, water balances can be calculated between times when the groundwater storage can be assumed to be at the same level (Dagg and Blackie, 1965, see discussion; Baugh, 1972).

2.4.5.2. Sediment Yield

Erosion, as measured by the amount of solid material being transported in the river as suspended sediment and bed-load, is notoriously difficult to measure accurately. No satisfactory method of measuring bed-load is available other than using sediment traps. Suspended sediment varies with discharge, and not only has a sediment rating curve to be built by sampling at specified discharges, but also the flow–frequency curve of the river has to be known to calculate the total yield of sediment over a given period (Alleyne and Percy, 1966; Piest, 1964). Plot experiments are very useful for determining the relationship between rainfall intensity and erosion (Kincaid *et al.*, 1966; Hudson, 1957; Ahmad and Breckner, 1974; Young and Niersma, 1973; Elwell and Stocking, 1973), but since plots can rarely sample the range of slopes, cultural practices, soil types and aspects encountered in the basic unit for water control planning—the catchment (Riggs, 1961; Eyre, 1965; Laycock, 1965)—better comparative measurements of regional erosion under different cultural practices are obtained from catchments. Within these, erosion plots can be used to give more detailed information on particular crops and cultivation methods.

In addition, erosion may be increased by the introduction of roads, tracks, embankments and settlements into forested areas and by the modification of drainage lines. These effects are associated with the development and intensification of agriculture but would not be measured by erosion plots on cultivated ground.

A new technique for measuring erosion has been suggested by Ritchie *et al.* (1974). It may be possible to calculate soil erosion from catchments by determining the distribution of the isotope ^{137}Cs in the soil.

2.4.5.3. Occult Precipitation

Some forests are known to act as collectors of occult precipitation by condensing atmospheric moisture onto cool surfaces of leaves (Geiger, 1965; Aubreville, 1949; Hori, 1953; Nagel, 1956). The effect is limited to areas where fog or low cloud are prevalent and will be offset by the increased rate of evaporation from the intercepted water.

In areas where fog drip or occult precipitation are thought to be a significant proportion of the total rainfall, i.e. greater than 5 per cent, analysis of the gross precipitation, throughfall and stemflow records should give estimates of the quantities involved.

2.4.5.4. Influence of Forest on Precipitation

It has been asserted that forests increase rainfall over the areas they occupy (Zon, 1927; Nicholson, 1929), with the corollary that removing the forests can effect the water resources of an area by decreasing the rainfall. Penman (1963) discusses the meterological argument and concludes with Kittredge (1948) that there is no reliable evidence of forests having more than a marginal influence on precipitation.

One of the reasons for the continuation of this controversy has been the difficulty of measuring rainfall in a forest. Gauges in clearings are sheltered from strong winds, and it is considered that this can lead to an overestimation of rainfall by up to 10 per cent compared with open ground (Kittredge, 1948, p. 98). Recent work on the effects of exposure on raingauges (Struzer *et al.*, 1968; Rodda, 1967) would suggest that raingauges in open ground tend to underestimate rainfall if not protected from the effect of wind, and that if both the above effects are taken into account, precipitation amounts should be more or less identical inside and outside a forested area.

2.4.6. STREAMFLOW MEASUREMENT

Installation of weirs or flumes to measure streamflow represents the most expensive capital item in catchment instrumentation. At the outset, it should be decided whether calibrated river sections, with their lower accuracy but much lower cost, are adequate to fulfill the research objectives or whether a weir or flume is required. Given that peak discharges are important in erosion studies, any structure designed for the catchment must be capable of covering an adequate range of discharges. Compound or multiple structures may cover the desired range at a lower cost but they seldom have satisfactory theoretical ratings and calibration by current meter or dilution gauging will be necessary (British Standard, 1964).

Several excellent design manuals are available for hydraulic structures (WMO 1971, which incorporates British Standard 3680 Parts 4A and 4B; Ackers and

Harrison, 1963; Parshall, 1953). While critical depth flumes are preferable to weirs in silt-laden waters, their cost is prohibitive in large rivers. When weirs are used, however, the effect of the reduced water velocities on sediment depositing upstream of the structures must be taken into account in assessing sediment yield.

2.4.7. SOME RECORDED RESULTS OF CATCHMENT STUDIES

Although extrapolation of experimental results obtained under different conditions of climate, soils and vegetation can be misleading, it is useful to consider the magnitude and type of result obtained elsewhere as an aid to experimental design.

The effects of forest clearing in temperate latitudes vary from insignificant increases in annual streamflow to 52 per cent (Coweeta 17, first year) depending on the percentage of the catchment cleared and the type of treatment immediately following the cutting (Hibbert, 1967). Nakano (1967) reports from Japan that removal of forest vegetation increased annual streamflow by 8 to 24 per cent.

Afforestation, on the other hand, has resulted in decreases in streamflow in the US (Schneider and Ayer, 1961; Eschner, 1965), and Wicht (1967) has recorded a 26 per cent decrease in streamflow over a period of 16 years in a forested catchment relative to a control in South Africa.

In the tropics replacing indigenous forest by alternative forms of land use may produce much smaller increases in streamflow depending on the type of vegetation chosen. Plantation forest and tea were found to have small effects on streamflow in East Africa once the plants completely covered the ground and had established deep root systems (Dagg and Blackie, 1965; Blackie, 1972b). Results from the EAAFRO experiment in Kenya show that clearing indigenous bamboo forest except for protective strips along the streambanks at Kimakia resulted in an increase of 16 per cent in annual streamflow compared with the control.

Annual crops such as maize and vegetables, however, use considerably less water than indigenous forest, and at Mbeya in Tanzania, where a six-month dry season restricts the period of cultivation, long-term average increases in streamflow of the order of 50 per cent have been recorded compared with the forested control (Edwards and Blackie, 1975).

With regard to accelerated erosion following forest cutting in the US, tenfold increases in sediment yield have been recorded by McClurkin (1967) and threefold increases by Anderson (1954) including the effects of logging and road-building—industrial rainstorms immediately following disturbance due to road construction can produce enormous increases in sediment yield, for example, 250 times the sediment yield of an undisturbed adjacent catchment (Fredriksen, 1965). On the other hand, under certain conditions, logging itself may have no measurable effect on erosion (Blackmarr and White, 1964; Brown and Krygier, 1974).

Perhaps the most dramatic demonstration of the ability of forests to decrease flood peaks and erosion losses can be found in the history of the Tennessee Valley

Authority (TVA, 1962). While the afforestation was highly successful in reducing flood peaks by 90 per cent and sediment loads by 96 per cent, the total yield of water from the forested catchments was almost halved (Pereira, 1973).

Plot experiments such as Hudson's (1957) in Rhodesia, Othieno's (1975) in Kenya, and Lal's in Nigeria (Chapter 4.3) illustrate how important crop type, treatments, slope and percentage ground cover are in determining the amount of soil loss from cultivated ground. Coulter (1972) draws attention also to the beneficial effects of establishing ground cover immediately following forest clearing in preventing soil erosion during the development of new rubber plantations in Malaya.

2.4.8. CONCLUSIONS

There is undoubtedly a pressing need to rationalize the experience gained from existing catchment experiments in order to produce general conclusions with regional application. At the same time, however, there are still considerable gaps in knowledge of the physical processes in certain problem areas such as the humid tropics, and further collection of basic data is needed.

While some planning decisions can be taken in the light of existing experimental evidence, controversial issues, such as the need for imposing unpopular erosion control measures, call for unambiguous scientific evidence which can often only be obtained in the location concerned.

In view of the increasing importance of conserving soil and water resources in the face of rapid population growth, there is an obligation to use every means at our disposal to establish guidelines for future agricultural development. Catchment research, although more costly and time-consuming than many other types of agricultural research, offers one of the few ways of simplifying the environmental processes and yielding quantitative results which can be readily interpreted in terms of land use planning.

REFERENCES

Ackermann, W. C., 1966, Guidelines for research on hydrology of small watersheds, US Dept. Interior, OWRR, Washington DC.

Ackers, P., and A. J. M. Harrison, 1963, Critical depth flumes for flow measurement in open channels, Hydraulics Research Paper 5, HMSO, London.

Ahmad, N., and E. Breckner, 1974, Soil erosion on three Tobago soils, *Trop. Agric. Trin.*, **51** (2), 313–324.

Alleyne, E. P., and M. J. Percy, 1966, Runoff and soil loss on two small watersheds in the Northern Range, Trinidad, *Trop. Agric. Trin.*, **43** (4), 323–326.

American Geophysical Union, 1965, Inventory of representative and experimental watershed studies conducted in the United States, Am. Geophys. Union.

Anderson, H. W., 1954, Suspended sediment discharge as related to streamflow, topography, soil and land-use, *Trans. Am. Geophys-Union*, **30**, 268–281.

Aubreville, A., 1949, *Climats, Forêts et Désertification de L'Afrique Tropicale*, Société d'Editions Geographiques, Maritime et Coloniales, Paris.

Baugh, J. R., 1972, Base flow recession as an index of representativeness in the hydrological regions, *IAHS/UNESCO Proc. Wellington Symp. Dec. 1970*, Vol. I, pp. 602–613.

Bell, J. P., 1969, A new design principle for neutron soil moisture gauges: the Wallingford Neutron Probe, *Soil Sci.*, **108**, 160–164.

Blackie, J. R., 1972a, The application of a conceptual model to two East African catchments, Unpubl. M.Sc. Thesis, Imperial College, London.

Blackie, J. R., 1972b, Hydrological effects of a change in land-use from rain forest to tea plantation in Kenya, *UNESCO/IAHS Proc. Wellington Symp. Dec. 1970*, Vol. 2, 312–329.

Blackmarr, W. H., and D. P. White, 1964, A 22-year progress report on hydrologic studies on the Rose Lake watershed in southern Michigan, *Mich. Agr. Exp. Stat. Quart. Bull.*, **46**, 542–560.

Boughton, W. C., 1967, Plots for evaluating the catchment characteristics affecting soil loss—1. Design of experiments, *J. Hydrol.* (*N.Z.*), **6**, 113–119.

British Standard 3680, Part 2A, 1964, Methods of measurement of liquid flow in open channels: Dilution methods: Constant rate injection, British Standards Institution, London.

Brown, G. W., and J. T. Krygier, 1971, Clearout logging and sediment production in the Oregon Coast Range, *Wat. Resources Research*, **7**, 1189–1198.

Chidley, T. R. E., and K. M. Keys, 1970, A rapid method of computing areal rainfall, *J. Hydrol.*, **12**, 15–24.

Clarke, R. T., and K. A. Edwards, 1973, The application of the analysis of variance to mean areal rainfall estimation, *J. Hydrol.*, **15**, 97–112.

Coulter, J. K., 1972, Soils of Malaysia, *Soils and Fert.*, **35**, 475–498.

Da Costa, J. A., and J. Jacquet, 1965, Presentation des resultats de l'equete UNESCO/AIHS sur les bassins representatifs et experimentaux dans le monde, *Bull. IAHS*, **10**, 107–119.

Dagg, M., and J. R. Blackie, 1965, Studies on the effects of changes in land use on the hydrological cycle in East Africa by means of experimental catchment areas, *Bull. IAHS*, **10**, 63–75.

Edwards, K. A., 1972, Estimating areal rainfall by fitting surfaces to irregularly spaced data, *WMO Symp. on Precip. in Mountainous Regions, Geilo*, Vol. 2, pp. 565–587.

Edwards, K. A., and J. R. Blackie, 1975, Hydrological research in East Africa, *E.A. agric. for. J.*, Special Issue.

Elwell, H. A., and M. A. Stocking, 1973, Rainfall parameters for soil loss estimation in a sub-tropical climate, *J. agric. eng. Res.*, **18**, 169–177.

Engler, A., 1919, Einfluss des Waldes auf den stand der Gewasser (Sperbel und Rappen, 1903–1915), *Mitt. Schwein Anst. Forstl. Versuchsw.*, **12**.

Eschner, A. R., 1965, Forest protection and streamflow from an Adirondack watershed, Unpubl. Ph.D. Thesis, State College of Forestry, Syracuse, NY.

Eyre, A., 1965, Watersheds and resource conservation, *Information Bull. of Sci. Res. Council Jamaica*, **6**, 65–73.

Fredriksen, R. L., 1965, Sedimentation after logging road construction in a small western Oregon watershed, *Proc. Fed. Inter-Agency Sedimentation Conf. 1963*, USDA Misc. Pub. 1970, Washington DC, pp. 56–59.

Geiger, R., 1965, *The Climate Near the Ground*, Harvard Univ. Press, Cambridge, Mass.

Hamilton, E. L., and P. B. Rowe, 1949, *Rainfall Interception by Chaparral in California*, Calif. For. Range Exp. St., USDA For. Service, Calif. State.

Herbst, P. H., and E. M. Shaw, 1969, Determining rain gauge densities in England from limited data to give a required precision for monthly rainfall estimates. *J. Wat. Engng*, **23**, 210–229.

Hewlett, J. D., and A. R. Hibbert, 1961, Increases in water yield after several types of forest cutting, *Bull. IAHS*, **6**, 5–17.
Hewlett, J. D., R. W. Lull, and K. G. Reinhart, 1969, In defense of experimental watersheds, *Wat. Resources Res.*, **5**, 306–316.
Hibbert, A. R., 1967, Forest treatment effects on water yield, In W. E. Sopper and H. W. Lull (Eds.), *Forest Hydrology*, Pergamon, Oxford, pp. 527–543.
Hirata, T., 1929, Contributions to the problem of the relation between the forest and water in Japan, *Imp. For. Sba.*, Megur, Tokyo.
Hori, T. (Ed.), 1953, *Studies on Fogs in Relation to Fog-preventing Forest*, Tonne Trading Co., Sapporo, Hokkaido, Japan.
Hudson, N. W., 1957, Erosion control research, *Rhodesia Agric. J.*, **54**, 297–323.
Huff, E. A., and J. C. McNeil, 1957, Rainfall relations on small areas in Illinois, *Ill. State Wat. Surv. Bull.*, **44**.
Hutchinson, P., and W. J. Walley, 1972, Calculation of areal rainfall using finite element techniques with altitudinal corrections, *Bull. IAHS.*, **17**, 259–272.
Ibbitt, R. P., 1974, Representative data sets for comparative testing of mathematical models for representative and experimental basins, UNESCO, 1st Prel. Ed.
IAHS, 1959, Symposium on Water and Forests, Hannoversh-Munden, Pub. 48.
IAHS, 1962, Colloquium on continental erosion, Bari, Pub. 59.
IAHS, 1965a, Symposium on Representative and Experimental Areas, Budapest, Pub. 66.
IAHS, 1965b, Symposium on design of hydrometeorological networks, Quebec, Pubs. 67, 68.
IAHS, 1971, Symposium on the World Water Balance, Reading, Pubs. 92, 93 and 94.
Kincaid, D. R., H. D. Osborne, and J. H. Gardner, 1966, Use of unit-source watersheds for hydrologic investigations in the semi-arid SW. *Wat. Resources Res.*, **2**, 381–392.
Kittredge, J., 1948, *Forest Influences*, McGraw-Hill, New York.
Laycock, A. H., 1965, The need for a regional approach in watershed management planning, *Amer. Water. Resources Ass. Proc.*, **1**, 146–154.
Lee, P. S., P. P. Lynn, and E. M. Shaw, 1974, Comparison of multiquadratic surfaces for the estimation of areal rainfall, *Bull. IAHS*, **19**, 303–317.
Leopold, L. B., 1972, Hydrologic research on instrumented watersheds, *UNESCO/IAHS Proc. Wellington Symp.*, Vol. 1, pp. 135–150.
Liebermann, J. A., and M. D. Hoover, 1951, Streamflow frequency changes on the Coweeta experimental watersheds, *Trans. Am. Geophys. Union*, **32**, 73–76.
Lull, H. W., and K. G. Reinhart, 1967, Increasing water yield in the north west by management of forested watersheds, US Forest Serv. Res. Pap. NE-66.
Mandeville, A. N., and J. C. Rodda, 1970, A contribution to the objective assessment of areal rainfall amounts, *J. Hydrol. (NZ)*, **9**, 281–291.
Martinelli, M., 1967, Watershed management in the Rocky Mountain alpine and subalpine zones, U.S. Forest Serv. Res. Note RM-36.
McClurkin, D. C., 1967, Vegetation for erosion control in the southern coastal plain of the United States, In W. E. Sopper and H. W. Lull (Eds.), *Forest Hydrology*, Pergamon, Oxford, pp. 655–661.
McCulloch, J. S. G., 1961, Statistical assessment of rainfall, *CCTA Inter. African Conf. on Hydrol. Nairobi*, CCTA/CSA PUB. 66, London, pp. 108–114.
McGuiness, J. L., 1963, Accuracy of estimating watershed mean rainfall, *J. Geophys. Res.*, **68**, 4763–4767.
Nagel, J. F., 1956, Fog precipitation on Table Mountain, *Quart. J. Roy. Met. Soc.*, **82**, 452–460.
Nakano, H., 1967, Effects of changes in forest conditions on water yield, peak flow and direct runoff of small watersheds in Japan, In W. E. Sopper and H. W. Lull (Eds.), *Forest Hydrology*, Pergamon, Oxford, pp. 557–564.
Nicholson, J. W., 1929, The influence of forests on climate and water supply in Kenya, Kenya Colony Forest Dept. Pamphlet 2.

Nicks, A. D., 1965, Field evaluation of raingauge network design principles, *WMO/IAHS. Symp. Quebec*, IAHS Pub. 67, pp. 82–93.
Othieno, C. O., 1975, An assessment of soil erosion on a field of tea under different soil management, In K. A. Edwards and J. R. Blackie, Hydrological research in East Africa, *E.A. agric. for. J.*, Special Issue.
Pacific Southwest Inter-Agency Committee, 1962, Limitations in hydrologic data as applied to studies of water control and water management, Hydrology Subcommittee.
Panchang, G. M., and R. Narayan, 1962, Adequate numbers of raingauges for accurate estimation of mean depths, *Irrig. and Power*, **19**, 94–107.
Parshall, R. L., 1953, Parshall flumes of large size, *Colorado Agric. Expt. Stat. Bull.*, **426** A.
Penman, H. L., 1948, Natural evaporation from open water, bare soil and grass, *Proc. Roy. Soc. London (A)*, **193**, 120–145.
Penman, H. L., 1963, Vegetation and hydrology, Tech. Comm. No. 53, Comm. Bureau of Soils, Harpenden.
Pereira, H. C., 1973, *Land-use and Water Resources*, Cambridge University Press, London.
Pereira, H. C., *et al.*, 1962, Hydrological effects of changes in land-use in some East African catchment areas, *E.A. agric. for. J.*, **27**, Special Issue.
Piest, R. F., 1964, Long-term sediment yields from small watersheds, *Proc. IAHS Gen. Ass. Berkeley*, Publ. 65, pp. 121–190.
Poncelet, L., 1959, Comparison of raingauges, *WMO Bull.*, **8**, 186–190.
Rakhmanov, V. V., 1966, *Rôle of Forests in Water Conservation*, translated by the Israel Program for Scientific Translations, Jerusalem.
Reinhart, K. G., and A. R. Eschner, 1962, Effects on streamflow of four different forest practices in the Allegheny Mountains, *J. Geophys. Res.*, **67**, 2433–2445.
Reynolds, E. R. C., and L. Leyton, 1963, *The Water Relations of Plants*, Blackwell Scientific Publications, Oxford, pp. 127–141.
Reynolds, E. R. C., and L. Leyton, 1967, Research data for forest policy: the purpose, methods and progress of forest hydrology, *Proc. 9th Brit. Common. For. Conf.*, Common. For. Inst., Oxford.
Riggs, F. E., 1961, The watershed as an entity for planning, In Tolley and Riggs, *Economics of Watershed Planning*, Iowa State Univ. Press., pp. 59–69.
Ritchie, J. C., J. A. Spraberry and J. R. McHenry, 1974, Estimating soil erosion from the redistribution of fallout ^{137}Cs, *Soil Sci. Soc. Am. Proc.*, **38**, 137–139.
Rodda, J. C., 1967, The systematic error in rainfall measurement, *J. Inst. Wat. Engng*, **21**, 173–177.
Rodier, J., 1959, Résultats obtenus sur les bassins expérimentaux de la France d'outre Mer, *La Houille Blanche, Grenoble*, **14** (Spec. No. 13), 952–963.
Rodier, J., and C. Auvray, 1965, Premiers essais d'étude générale du ruissellement sur les bassins expérimentaux et réprésentatifs d'Afrique tropicale, *IAHS. Symp. of Budapest*, Vol. I, pp. 12–38.
Rodier, J., and G. Vuillaume, 1972, Interprétation des résultats sur quelques bassins réprésentatifs tropicaux, *UNESCO/IAHS Wellington Symposium, Dec. 1970*, Vol. I, pp. 443–454.
Rutter, A. J., 1963, Studies on the water relations of *Pinus sylvestris* in plantation conditions, I. Measurements of rainfall and interception, *J. Ecol.*, **51**, 191–203.
Rycroft, H. B., 1949, Random sampling of rainfall, *J. S.A. Forestry Assoc.*, **18**, 71–81.
Salter, M. de Carle S., 1919, Exposure of raingauges, *British Rainfall*, **59**, 263–268.
Schneider, W. J., and G. R. Ayer, 1961, Effects of reforestation on streamflow in Central New York, USGS Wat. Supp. Paper 1602.
Sopper, W. E., and H. W. Lull, (Eds.), 1967, *International Symposium on Forest Hydrology*, Proc. Nat. Sci. Found. Adv. Sci. Seminar, Pennsylvania, 1965, Pergamon, Oxford.
Struzer, L. R., T. V. D'jackova, V. F. Kulagin and A. S. Zoycev, 1968, On errors in the

measurement of precipitation by raingauges at ground level in the presence of wind, *Glav. Geof. Obs. Trudy vyp.*, **215**, 135–152.

Tennessee Valley Authority, 1962, Reforestation and erosion control influences upon the hydrology of the Pine Tree Branch watershed, 1941–1960, TVA, Knoxville, Tenn.

Thornthwaite, C. W., 1948, An approach towards a rational classification of climate, *Geogr. Review*, **38**, 85–94.

Toebes, C., and V. Ouryvaev, (undated), Representative and experimental basins: an international guide for research and practice, UNESCO, *Studies and Reports in Hydrol.*, **4**.

UNESCO, 1971, Mathematical models in hydrology, *Proc. Warsaw Symp. Studies and Reports in Hydrol.*, **15**.

UNESCO/IAHS, 1972, Results of research on representative and experimental basins, *Proc. Wellington Symp. Dec. 1970*, Two Volumes, IAHS/UNESCO.

UNESCO/WMO, 1974, *Records of the Int. Conf. on the Results of the Intern. Hydrol. Decade and on Future Programmes in Hydrology, Sept. 1974*, Vol. I, Final Report.

Ward, R. C., 1971, Small watershed experiments: An appraisal of concepts and research developments, Univ. Hull, Occasional Paper 18.

Wicht, C. L., 1967, Conclusions from South African multiple watershed experiments, In W. E. Sopper and H. W. Lull (Eds.), *Forest Hydrology*, Pergamon, Oxford, pp. 749–759.

WMO, 1971, Use of weirs and flumes in stream gauging, Tech. Note 117, World Met. Org. No. 280.

Wolf, P. O., 1966, Comparison of methods of flood estimation, In *River Flood Hydrology*, Proc. Symp. Inst. Civ. Eng. March 1965, Inst. Civ. Eng., London, pp. 1–25.

Young, R. A., and J. L. Niersma, 1973, The rôle of rainfall impact in soil detachment and transport, *Wat. Resources Res.*, **9** (6), 1629–1636.

Zon, R., 1927, Forests and water in the light of scientific investigation, U.S. Nat. Waterways Comm. Final Rep., App. V, 62nd Congress, 2nd Sess., Sen-Doc, pp. 205–302.

PART 3

Characterization of Soils and Climates in Relation to the Erosion Hazard

3.1

Analysis of Factors Affecting Rainfall Erosivity and Soil Erodibility

R. LAL

The terms 'erosivity' and 'erodibility' have been defined in Chapter 2.1. While erosivity refers to the aggressiveness of the climate, the term erodibility refers to the liability of the soil to suffer erosion due to the forces causing detachment and transport of soil particles. The analysis of both is essential to an understanding of the erosion process, its importance in the humid tropics, and the development of measures for the control of erosion.

3.1.1. EROSIVITY

Tropical rains are more erosive than temperate rains, becauseof the high intensity of tropical storms (Table 3.1). A majority of the tropical rainstorms fall in the category of 'erosive' storms, while only 5 to 10 per cent of temperate storms are erosive. Tropical storms are generally accompanied by high-intensity winds which further increase their aggressiveness. The kinetic energy of a storm is affected by various factors, some of which are briefly described below.

Table 3.1 Some examples of high intensity rainfall from tropical regions (after Jennings, 1950)

Duration	Rainfall (cm)	Location	Date
15 min	4·06	Ibadan, Nigeria	June 16, 1972
15 min	11·94	Monrovia, Liberia	Aug. 1, 1974
15 min	19·81	Plumb Point, Jamaica	May 12, 1916
90 min	25·4	Colombo, Sri Lanka	1907
24 hr	50·8	Kalani Valley, Sri Lanka	May, 1940
24 hr	116·81	Baguio, Philippines	July 14–15, 1911
48 hr	167·10	Funkiko, Formosa	July 19–24, 1913
63 hr	200·96	Baguio, Philippines	July 14–17, 1911
96 hr	258·68	Cherrapunji, India	June 12–15, 1876

3.1.1.1. Factors Affecting Rainstorm Erosivity

Rainfall intensity

The initiation of erosion by rainfall is due to detachment of a soil particle from the bulk soil. The detached particle is then removed by transport in run-off water (Chapter 2.2). The force causing detachment is that associated with the impact of the individual water drop. The question of whether this is due to the momentum or kinetic energy of the drop is still a matter of debate. This discussion is given in terms of kinetic energy ($\frac{1}{2}mv^2$) rather than momentum (mv) as most published work is based on the assumption that kinetic energy of rain is the causative factor in initiating detachment and erosion. Because the energy of the individual drop is related to its size, the energy of an individual storm is related to its drop size distribution. The total energy of any rainstorm is the sum of its component drops, and so there is a relation between total energy and amount of rain, and, empirically, with the intensity of any storm, as described by Kowal and Kassam in the following chapter. Tamhane *et al.* (1959) found an exponential relation between runoff and intensity and soil loss and intensity.

The intensity has also been closely related to total kinetic energy by Wischmeier and Smith (1958), as shown in equation (3.1):

$$E_k = 916 + 331 \log_{10} I \tag{3.1}$$

in which E_k is kinetic energy in foot-tons per acre–inch of rain, and I is intensity in inches per hour. In metric units this equation becomes

$$E_k = 210{\cdot}3 + 89 \log_{10} I \tag{3.2}$$

where E_k is in metre-tonnes per hectare per cm of rain, and I is rainfall intensity in cm per hour.

Drop Size Distribution

Raindrop size, distribution, and shape all influence the energy, momentum and erosivity of a rainstorm. Laws and Parsons (1943) reported an increase in median drop size with increase in rain intensity. The relation between median drop size (D_{50} in mm) and rainfall intensity they found is:

$$D_{50} = 2{\cdot}23\ \mathrm{I}^{0{\cdot}182} \tag{3.3}$$

where I is intensity in inches per hour. The maximum drop diameter of natural rainfall is probably close to 7 mm. Hudson (1971) pointed out that for tropical rainstorms, a relationship of the type $D_{50} \times I^b$ is valid only for low-intensity ranges, and so may not apply to erosive storms in the humid tropics.

Terminal Velocity

The kinetic energy of a rainstorm is related to the velocity of the raindrops at the time of impact with the soil. The distance through which a raindrop must fall before attaining its terminal velocity is a function of drop size distribution (Table 3.2).

Table 3.2 Relationship between drop size and terminal velocity (Wischmeier and Smith, 1960)

Drop size (mm)	Terminal velocity (m/sec)	Fall to reach 95% terminal velocity (m)
0·25	1·0	–
0·50	2·0	–
1·00	4·0	2·2
2·00	6·5	5·0
3·00	8·1	7·2
4·00	8·8	7·8
5·00	9·1	7·6
6·00	9·3	7·2

The kinetic energy of a rainstorm is related to terminal velocity according to equation (3.7) and as shown in Table 3.3:

$$E_k = \frac{IV^2}{2} \qquad (3.4)$$

where E_k is the energy in watts/m^2, I is the intensity in mm/sec and V is the velocity of rainfall before impact in m/sec.

Ellison (1947) developed the relationship between soil detached (splashed) and terminal velocity, drop diameter and rainfall intensity (equation 3.5):

Table 3.3. Kinetic energy of rainfall in relation to medium drop diameter and areal concentration of drops for rainfall of various intensities (after Lull, 1959)

Rainfall	Intensity (cm/hr)	Median diameter (mm)	Fall velocity (m/sec)	Number of drops (m^{-2}/sec)	Kinetic energy/m^2 (w/m^2)
Fog	0·013	0·01	0·003	67,000,000	$1{\cdot}6 \times 10^{-10}$
Mist	0·5	0·10	0·021	27,000	$3{\cdot}3 \times 10^{-9}$
Drizzle	0·025	0·96	4·1	150	$6{\cdot}1 \times 10^{-4}$
Light rain	0·10	1·24	4·8	280	$3{\cdot}3 \times 10^{-3}$
Moderate rain	0·38	1·60	5·7	495	$1{\cdot}7 \times 10^{-2}$
Heavy rain	1·5	2·05	6·7	495	$9{\cdot}4 \times 10^{-2}$
Excessive rain	4·1	2·40	7·3	820	$8{\cdot}9 \times 10^{-1}$
Cloudburst	10·0	2·85	7·9	1,214	$9{\cdot}2 \times 10^{-1}$
Cloudburst	10·0	4·00	8·9	440	1·1
Cloudburst	10·0	6·00	9·3	130	1·2

$$E = KV^{4\cdot33}d^{1\cdot07}I^{0\cdot65} \tag{3.5}$$

where E is the relative amount of soil detached, K is a constant of the soil, V is the velocity of raindrops in feet per second, d is the diameter of raindrops in millimetres and I is rainfall intensity in inches per hour.

Wind Velocity

Wind velocity accompanying a storm influences its kinetic energy and hence the erosive capacity. Wind is an important characteristic of tropical rainstorms (Ahmad and Breckner, 1974). Lyles *et al.* (1969) reported that wind-driven rain was more effective in clod-breaking than anticipated. Up to 73 per cent more soil detachment occurred at 13·4 m/sec wind velocity than for no wind (Table 3.4).

Table 3.4. Effect of wind velocity on soil detachment (after Lyles *et al.*, 1969)

Wind velocity (m/sec)	Soil detachment (arbitrary units) at different intensities (cm/hr) 1·6	2·84	5·61
0	56	93	97
6·7	95	98	100
13·4	97	100	100

Direction of the Slope

Slope direction in relation to the direction of the storm can significantly alter the actual kinetic energy of the drop impact on soil because of the angle at which it hits the soil particle. The maximum impact will be for normal (90° angle) impact.

3.1.1.2. Erosivity Indices

(a) EI_{30}. Experiments by various researchers have indicated that neither the rainfall amount nor the intensity are very highly correlated with soil loss. Though kinetic energy of the rain is generally accepted to be the driving force, it is a difficult parameter to monitor. Various attempts have been made to determine a combination of different rainfall parameters which would be a better index of erositivity than either the amount or intensity. Wischmeier (1955) suggested that a combination of rainfall energy and quantity was likely to provide a close correlation with soil erosion, and analysis of much field evidence by Wischmeier *et al.* (1958) showed that the correlation of soil loss with amount of rain in individual storms, as well as with maximum amount falling in 5–, 15– or 30–minute intervals, was poor. However the product of the kinetic energy of the storm and the maximum intensity of the rainfall during the first 30 minutes of a storm (EI_{30}) was most significantly correlated with soil loss determined on standard field plots. This relation has subsequently been extensively used in predicting soil loss (Hudson, 1971 and Chapter 5.4). Some modifications and adjustments in the EI_{30} index have been proposed by Wischmeier (1974).

(b) $KE > 1$. Experiments conducted in Rhodesia by Hudson (1971) indicated that accumulative kinetic energy of storms with intensity greater than 2·5 cm/hr was better correlated with soil loss than the EI_{30} index. Hudson pointed out that storms with intensity of less than 2·5 cm/hr were not erosive.

(c) p^2/P. Fournier (1956) and Fournier and Henin (1959) analysed the sediment load values of more than 140 rivers in Europe, Asia and the US and found a significant correlation between a rainfall coefficient

$$\frac{p^2}{P} = C \tag{3.6}$$

where p is the mean monthly rainfall in the wettest month of the year and P is the mean annual rainfall, and the soil loss or 'specific degradation' (D) in tonnes/km^2 of the catchment area of the river

$$D = aC - b \tag{3.7}$$

where a and b are coefficients whose magnitude depends upon the orographic coefficient O of equation (3.8):

$$O = \overline{H}_2/S \tag{3.8}$$

$\overline{H}$ is the mean height of the terrain above its base level and S is its projected area.

This climatic index was more significantly correlated with erosion in the tropics than in the temperate regions and with steeper than with gentler slopes, but the fact that it refers to catchment areas with variable land use limits its applicability in relation to soil management.

(d) AI_m. Experiments conducted in the tropics have indicated a lower correlation coefficient between EI_{30} and soil loss (Ahmad and Breckner, 1974; Lal, 1976 a, b) than was obtained for the original experiments in USA. Lal (1976a, b) reported a better correlation with the product of total rainfall amount (A) and peak storm intensity (I_m) than either EI_{30} or $KE > 1$. The annual index AI_M is calculated according to equation (3.9):

$$AI_m = \{\sum_{1}^{12}\sum_{1}^{n}(ai_m)\} \tag{3.9}$$

where a is the total rainfall in any one storm in cms and i_m is the maximum storm intensity in cms/hr and n is the number of rainy days in the month.

The weighted mean average correlation coefficient of runoff and soil loss with various erosivity indices is shown in Table 3.5 (Lal, 1976a, b).

Table 3.5. Weighted mean average correlation coefficient (r) of various erosivity indices with runoff and soil loss for standard nonvegetated runoff plots at IITA, Ibadan, Nigeria, with slopes 1, 5, 10 and 15 per cent and for all years 1971 to 1975 inclusive (full details and complete data for runoff and soil loss from each plot are given in Lal (1976c).)

	Correlation coefficient (r) for runoff and soil loss	
R	Run off	Soil loss
$KE > 1$	0·81	0·64
EI_{30}	0·85	0·65
AI_{m}	0·91	0·80

3.1.2. ERODIBILITY

The different erosion of two soils under similar environmental and management conditions is attributed to their inherent soil characteristics. This property of the soil is referred to as soil erodibility. Some of the problems involved in studying erodibility are mentioned in Chapter 2.1. Erodibility involves those soil properties that affect the infiltration rate and permeability, and the changes with time that occur in those soil properties, and others that determine the effects of the dispersion, splashing, abrasion and transporting forces of rainfall and runoff. Structure and structural stability (Chapter 2.2) are closely related to these soil properties, but are difficult to define in such a way that they provide a measure of erodibility. Empirical determinations of erodibility as defined by the Universal Soil Loss Equation (Chapters 3.3 and 5.4) are likely to be of greatest general use.

3.1.2.1. Detachability and Transportability of Soil

Various soil properties affect its detachability: for instance, particle size distribution, organic matter content, presence of cementing materials such as Fe and Al oxides, nature of clay minerals and balance of cations on the exchange complex, and properties which are themselves dependent on these such as permeability, soil structure and strength. Wischmeier *et al.* (1969) developed a soil erodibility nomogram based on these soil properties. Some other properties such as entrapped air may also be important. The structural stability of a soil largely determines detachability, and is discussed more fully in Chapter 2.2. Soil transportability depends both on soil and water flow characteristics. For instance, clay sized particles are more easily transported than fine sand.

Flow characteristics significantly influence the transportability. Little (1940) and Ellison (1947) defined transportability. Ellison (1947) related the transportability (D_3) of the stream flow to the following characteristics (equation 3.10):

$$D_3 = f\{(V^2/2g)\beta M\} \tag{3.10}$$

where $V^2/2g$ is the energy gradient of flow, M is the quantity of abrasive material per unit of flow, and β is a measure of the flow characteristics and transportability as defined by Ekern (1954).

REFERENCES

Adams, J. E., 1957, A rainfall simulator and erodibility of some Iowa soils, *Iowa St. Coll. J. Sci.*, **31**, 347–348.

Ahmad, N., and E. Breckner, 1974, Soil erosion on three Tobago soils, *Trop. Agric. Trin.*, **51**, 313–324.

Best, A. C., 1950, The size distribution of raindrops, *Quart. J. Roy. met. Soc.*, **76**, 16–36.

Bruce Okine and R. Lal, 1975, Soil erodibility as determined by raindrop technique, *Soil Sci.*, **119**, 149.

Burns, R. V., 1947, Soil erosion in Ceylon with particular reference to floods, *Trop. Agricst. (Ceylon)*, **103**, 240.

Collis-George, N., and R. Lal, 1971, Infiltration and structural changes as influenced by initial moisture content, *Aust. J. Soil Res.*, **9**, 107–116.

Ekern, P. C., 1954, Rainfall intensity as a measure of storm erosivity, *Soil Sci. Soc. Am. Proc.*, **18**, 212–216.

Ekern, P. C., and R. J. Muckenhirn, 1947, Water drop impact as a force in transporting sand, *Soil Sci. Soc. Am. Proc.*, **12**, 441–444.

Ellison, W. D., 1944, Studies of raindrop erosion, *Agric. Eng.*, **25**, 131–136, 181–182.

Ellison, W. D., 1947, Soil erosion studies, Parts I, II, III, IV, V, VI, VII, *Agric. Eng.*, **28**, 145–146, 197–201, 245–248, 297–300.

Fournier, F., 1956, The effects of climatic factors on soil erosion. Estimates of solids transported in suspension in runoff, *Ass. Int. Hydrol. Publ.*, **38**, 6.

Fournier, F., and S. Henin, 1959, A new climatic formula for evaluating the specific degradation of soil, *C. R. Acad. Sci.*, **248**, 1694–1969.

Free, G. R. 1960, Erosion characteristics of rainfall, *Agric. Eng.* **41**, 447–449, 455.

Hudson, N., 1971, *Soil Conservation*, Batsford, London.

Jennings, A. H., 1950, Worlds greatest observed point rainfalls, *Month. Weather Rev.*, **78**, 4–5.

Lal, R., 1976a, Soil erosion on Alfisols in western Nigeria. I. Effect of slope, crop rotation, and residue management, *Geoderma*, **16**, 363–375.

Lal, R., 1976b, Soil erosion on Alfisols in western Nigeria. II. Effect of mulch rate, *Geoderma*, **16**, 377–387.

Lal, R., 1976c, Soil erosion on Alfisols in western Nigeria, *IITA Monograph No.* 1, IITA, Ibadan.

Laws, J. O., and D. A. Parsons, 1944, The relation of raindrop size to intensity, *Trans. Am. Geophys. Union*, **24**, 452.

Little, J. H., 1956, *Erosional Topography and Erosion*, Carlisle, San Francisco, Calif.

Lull, H. W., 1959, Soil compaction in forage and range lands, *USDA Misc. Publ.*, 768.

Lyles, L., L. A. Disrud and N. P. Woodruff, 1969, Effect of soil physical properties, rainfall characteristics and wind velocity on clod disintegration by stimulated rainfall, *Proc. Soil Sci. Soc. Am.*, **33**, 302–306.

Mihara, Y., 1959, Effects of raindrop and grass on soil erosion, *Proc. Sixth Int. Grassl. Congr.* 1953, pp. 987–990.

Rose, E. E., 1969, Soil detachment caused by rainfall, *Soil Sci.*, **89**, 28–35.

Tamhane, R. V., T. D. Biswas and B. Das, 1959, Effects of intensity of rainfall on soil loss, *J. Indian Soc. Soil Sci.*, **7**, 231–238.

Wischmeier, W. H., 1955, Punch cards record runoff and soil, *Agric. Eng.*, **36**, 664–666.

Wischmeier, W. H., 1974, New developments in estimating water erosion, *Proc. 29th Annual Meeting of the Soil Conservation Society of America, Aug.* 11–14, 1974, Syracuse, New York.

Wischmeier, W. H., and D. D. Smith, 1958, Rainfall energy and its relationship to soil loss, *Trans. Am. Geophys. Union*, **39**, 285–291.

Wischmeier, W. H., C. B. Johnson and B. C. Cross, 1969, A soil erodibility nomograph for farmland and construction sites, *J. Soil Water Conservation* **26**, 190.

Wischmeier, W. H., D. D. Smith and R. B. Uhland, 1958, Evaluation of factors in soil loss equation, *Agric. Eng.*, **39**, 458.
Woodburn, R., 1948, The effect of structural conditions on soil detachment by raindrop action, *Agric. Eng.*, **29**, 154–156.

3.2

Energy Load and Instantaneous Intensity of Rainstorms at Samaru, Northern Nigeria

J. M. KOWAL and A. H. KASSAM

3.2.1. INTRODUCTION

The energy load of individual rainstorm systems in the tropics is much greater than that of temperate and subtropical systems. This is consistently reflected in the concentrated nature of the precipitation in the tropics, which is both heavier and of higher intensity (Ellison, 1952; Hudson, 1971). Using high-speed photography, Ellison (1944) showed the erosive nature of raindrop splashes and the heavy battering effects of individual raindrops on the soil surface. These characteristics of tropical rainstorms result in the breaking up of soil aggregates, compaction and sealing of the soil surface, decrease in infiltration and increase in runoff and soil erosion.

At Samaru (11° 11′ N, 7° 38′ E, in the Northern Guinea Savannah zone, with an average annual rainfall of 1100 mm) individual storms with rainfall exceeding 50 mm and peak intensities of 120 to 160 mm/hr are not uncommon. Erosion studies conducted at Samaru by Kowal (1970a, b) and Kowal and Stockinger (1973) have shown that these rainstorms present serious erosion hazards and may result in a decline in soil fertility, particularly under continuous arable cultivation. In consequence, problems of soil, water and crop management arise that are largely unknown under the gentler rainfall of temperate and subtropical climates.

Conventional rainfall data are of limited value for development of sound practices of soil and crop management unless they are supplemented by an intimate knowledge of the characteristics of rainstorms which describe their erosiveness. In spite of their serious environmental and economic hazards, little quantitative information is available on the magnitude and the pattern of instantaneous intensity and the energy load of tropical rainstorms. In this chapter we present this kind of information for rainstorms at Samaru. The study is based on an analysis of graphical records of raindrops made during the wet seasons of

1973 and 1974, using an instrument previously described elsewhere (Kowal *et al.*, 1973).

3.2.2. METHODS

3.2.2.1. Instrument and Technique

The instrument and the accuracy of the technique used for the study were described in detail by Kowal *et al.* (1973). The instrument makes graphical records, on an adjustable time-scale, of the number and amplitude of electrical pulses originating from the impact of raindrops on a transducer surface, utilizing the piezoelectric effect. The amplitude of each pulse is recorded on a moving chart in the form of a vertical line whose length is proportional to the size of the intercepted drop. The spectrum of the total kinetic energy load and the drop size distribution of a particular rainstorm is calculated from the total number of pulses, their emplitudes and the total amount of rainfall as described by Kowal *et al.* (1973). Similarly, the instantaneous intensity and the kinetic energy load of a rainstorm for a particular time interval is calculated from the number and the amplitude of pulses corresponding to the various drop size classes.

In 1973, records were made using a flat transducer disc (40 mm in diameter), while in 1974 a convex sensor (chord length 36 mm, radius of curvature 36 mm) was used. The effective interception surface areas of these sensors were 12·56 and 10·18 cm^2 respectively. Changing to a curved intercepting sensor was found to reduce damping effects from a possible accumulation of water on the surface. Immediately before the rain, the sensitivity of the instrument was checked by means of standard drops falling from an aspirator supported on a stand.

Rainfall was measured with a standard 12·7 cm diameter raingauge. The transducer sensor for raindrop interception was placed 1 to 2 m away from the raingauge at the same elevation.

3.2.2.2. Records and Presentation

The records are analysed and presented in two parts. First the detailed analysis of one selected rainstorm is presented to indicate the nature of the various rainfall parameters such as total volume, drop size, drop number, intensity and energy load. The objects of this analysis were to provide an opportunity of testing the closeness of association between these parameters by means of simple correlations and to see if any simple measured parameter, for example rainfall amount, may be used reliably for assessing the energy load of rainstorms.

Next, an analysis of all the graphical records of rainstorms in 1973 and 1974 as well as the conventional rainfall data of the last 15 years are presented in terms of the magnitude and the seasonal distribution of the energy load of rainstorms. Complete graphical records of 22 rainstorms were made during 1973 and 1974. Of these, 18 rainstorms varied in amount from 6·60 to 60·64 mm. These were fully

Table 3.6. Number and size of drops and kinetic energy load distribution of a 20·07 mm rainstorm (rainfall volume = 2007 mm^3/cm^2) on August 21, 1973 at Samaru

Drop size class	Drop No. per cm^2	Drop No. (%)	Volume (mm^3/cm^2)	Average drop volume (mm^3)	Drop diameter (mm)	Kinetic energy (J/m^2)	Kinetic energy (%)
1	21·81	22·7	144	6·64	2·34	35	5·0
2	17·84	18·5	236	13·28	2·94	71	10·2
3	19·59	20·3	390	19·92	3·36	131	18·9
4	14·68	15·2	389	26·56	3·70	142	20·4
5	9·92	10·3	327	33·20	4·00	112	16·1
6	9·43	9·8	374	39·84	4·24	146	21·1
7	2·08	2·2	88	46·41	4·46	37	5·3
8	0·80	0·8	46·7	53·13	4·66	17	2·5
9	0·16	0·2	9·5	59·75	4·86	3·9	0·6
Total	96·31	100·0	2007	20·85	3·42	694	100·0

analysed. The remaining four rainstorms were drizzles ranging in amount from 0·89 to 3·05 mm. These were analysed for drop number only.

3.2.2.3. Symbols and Units

R_a = amount of rainfall in mm

R_v = volume of rainfall in mm^3/cm^2

R_i = rainfall intensity in mm/hr

D_n = drop number per cm^2

D_v = average drop volume in $mm^3 = R_v/D_n$

$$D_s = \text{average drop size in mm}^3 = \frac{R_v \text{ in a given drop size class}}{D_n \text{ in that size class}}$$

3.2.3. ANALYSIS OF A RAINSTORM IN DETAIL

3.2.3.1. Drop Size

A summary of the drop size distribution of a 20 mm, 18-minute duration rainstorm as typically experienced at Samaru is shown in Table 3.6.

The distribution of drop size (D_s) ranged from 6·6 to 59·8 mm^3 corresponding to a drop diameter range of 2·34 to 4·86 mm. The predominant drop size diameter was 2·34 mm, while the mean drop diameter for the rainstorm was 3·42 mm. This clearly shows that the drop size distribution of a short duration rainstorm at Samaru differs substantially from that of temperate and subtropical rainstorms, where the bulk of the rainfall is composed of much smaller drops (Bean and Wells, 1953; Hudson, 1971).

The drop size distribution of individual rainstorms varies considerably depending on the amount or volume, duration and intensity of rainfall. At Samaru, drizzles which occur infrequently are of prolonged duration, low volume and intensity, and composed of smaller drops. On the other hand, short duration rainstorms are of high volume and intensity, and composed of relatively larger drops. Figure 3.1 shows the drop size distribution of selected rainstorms at Samaru. The extent of the variation in drop size distribution between different types of rainstorms is clearly demonstrated. In particular, the greater the rainfall, the greater the proportion of larger drops.

A regression analysis of the amount of rainfall (R_a) in successive one-minute intervals (Table 3.7) on the corresponding average drop volume (D_v) (Table 3.8) showed a positive correlation ($r = 0{\cdot}73$) suggesting a tendency for the average drop size to increase with the amount or the rate of rainfall.

Table 3.7. Distribution of drop number m^{-2} per second, total drop number per minute and average drop volume (mm^3) in successive one-minute intervals

Drop size class	Time interval (min) 1	2	3	4	5	6	7	8	9	10	11	12	13	14	15	16	17	18	Average drop No. (m^{-2}/sec)
1	147	198	332	238	358	332	385	198	212	225	120	198	252	133	147	120	40	—	201·9
2	173	67	93	120	185	198	160	292	278	212	358	185	225	80	173	120	40	13	165·1
3	398	212	147	265	185	80	93	80	278	265	318	198	173	198	147	66	120	40	181·3
4	238	147	185	93	93	133	53	146	147	185	147	212	53	173	93	160	67	120	135·8
5	107	107	67	173	120	53	13	80	93	67	93	120	13	160	40	147	80	120	91·8
6	27	160	147	80	40	80	40	53	13	93	93	93	93	80	53	13	67	345	87·2
7	0	80	27	0	0	0	0	0	0	0	0	0	0	27	0	27	27	160	19·3
8	0	53	13	0	0	13	0	0	13	0	0	0	0	0	0	0	0	40	7·3
9	0	27	0	0	0	0	0	0	0	0	0	0	0	0	0	0	0	0	1·5
Total drop No. (m^{-2}/sec)	1090	1051	1011	969	981	889	744	849	1034	1047	1129	1006	809	851	653	653	441	838	891·4
Total drop No./cm^2/min	6·54	6·30	6·07	5·81	5·89	5·33	4·46	5·09	6·20	6·28	6·77	6·04	4·85	5·11	3·92	3·92	2·65	5·03	5·35
Average drop volume (mm^3)	20·31	26·97	20·90	20·52	16·89	17·60	13·44	18·22	18·27	19·52	20·04	20·91	16·94	23·76	19·56	22·36	24·86	36·81	20·85

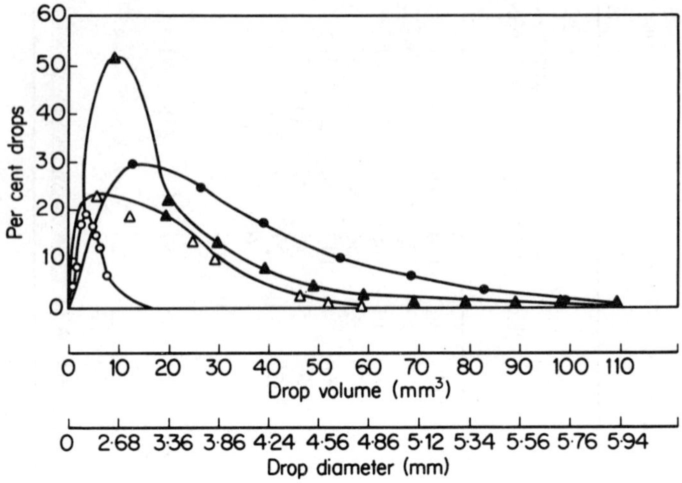

Figure 3.1. Drop size distribution of selected rainstorms at Samaru. △, typical rainfall (20·1 mm, 96 drops/cm²); ▲, composite (storm and drizzle) rainfall (6·0 mm, 528 drops/cm²); ○, drizzle (6·6 mm, 135 drops/cm²); ●, heavy rainfall (40·6 mm, 1228 drops/cm²)

3.2.3.2. Drop Number

The total number of drops (D_n) during the rainstorm was 96·31 per cm² (Table 3.6). This number is probably an underestimate of the actual number of drops since the recorder is not sensitive enough to register accurately very small drops (less than 1·5 mm diameter). However, this is of little consequence in estimating the energy load of tropical rainstorms (Kowal *et al.*, 1973). The number of drops per cm² intercepted in each successive one-minute interval (Table 3.7) ranged from 2·65 to 6·77 with a mean of 5·35.

There was a positive correlation ($r = 0{\cdot}64$) between the drop number (D_n) and the rainfall intensity (R_i) within the rainfall intensity range of 40 to 80 mm/hr. The regression coefficient was 0·05 (increase in drop number per cm²) for each 1 mm/hr increase in rainfall intensity. This suggests that about 41 per cent of the increase in the rainfall intensity was due to the increase in the number of drops. Of the remaining 59 per cent, 49 per cent could be accounted for (see below) by the increase in the size of drops. The unaccounted 10 per cent probably represents the error of the estimate.

3.2.3.3. Rainfall Intensity

The rainfall intensity during successive one-minute intervals (Table 3.8) ranged from 36·0 to 111·1 mm/hr. The average rate for the 18-minute period was 66·9 mm/hr.

Table 3.8. Distribution of rainfall volume (mm^3/cm^2), average amount of rainfall (mm) and average rainfall intensity (mm/hr) in successive one-minute intervals

Drop size class	Time interval (min) 1	2	3	4	5	6	7	8	9	10	11	12	13	14	15	16	17	18	Total volume (mm^3/cm^2)	Total volume (%)
1	5·8	7·9	13·2	9·5	14·3	13·2	15·3	7·9	8·4	9·0	4·8	7·9	10·0	5·3	5·8	4·8	1·6	0	144·8	7·2
2	13·8	5·3	6·6	9·6	14·7	15·8	12·7	23·2	22·2	16·9	28·5	14·7	17·9	6·4	13·8	9·6	3·2	1·1	236·1	11·8
3	47·6	25·3	17·5	31·7	22·1	9·6	11·2	9·6	33·3	31·7	38·0	23·7	20·7	23·7	17·5	8·0	14·3	4·8	390·2	19·4
4	38·0	23·4	29·5	14·9	14·9	21·5	8·5	23·4	23·4	29·5	23·4	33·7	8·5	27·6	14·9	25·5	10·6	19·1	389·9	19·4
5	21·2	21·2	13·3	34·5	23·9	10·6	2·7	15·9	18·6	13·3	18·6	23·9	2·7	32·9	8·0	29·2	15·9	22·2	327·7	16·3
6	6·4	38·0	35·1	19·1	9·6	12·7	9·6	12·7	3·2	22·3	22·3	22·3	22·3	19·1	12·7	3·2	15·9	81·1	374·1	18·6
7	0	22·3	7·4	0	0	0	0	0	0	0	0	0	0	7·4	0	7·4	4·3	44·1	92·9	4·6
8	0	17·0	4·2	0	0	4·2	0	0	4·2	0	0	0	0	0	0	0	0	12·7	42·5	2·1
9	0	9·6	0	0	0	0	0	0	0	0	0	0	0	0	0	0	0	0	9·6	0·6
Total volume (mm^3/cm^2)	132·9	170·0	126·0	119·2	99·5	93·8	60·0	92·8	113·3	122·6	135·6	126·3	82·1	121·4	72·8	87·6	65·9	185·1	2007·7	100·0
Average rainfall amount (mm)	1·33	1·70	1·27	1·19	0·99	0·94	0·60	0·93	1·13	1·22	1·36	1·26	0·82	1·21	0·73	0·88	0·66	1·85	20·07	
Average rainfall intensity (mm/hr)	79·7	102·0	76·1	71·6	59·7	56·3	36·0	55·7	68·0	73·5	81·4	75·8	49·3	72·9	43·7	52·6	39·5	111·1	66·92	

Table 3.9. Distribution of kinetic energy (KE) in erg/cm^2 and total kinetic energy in $Jm^{-2}\ mm^{-1}$ of rainfall in successive one-minute intervals

Drop size class	KE per drop (erg)	Time interval (min) 1	2	3	4	5	6	7	8	9	10	11	12	13	14	15	16	17	18	Total KE $J \times 10^{-7}$
1	1590	1399	1892	3164	2273	3418	3164	3672	1892	2019	2146	1145	1892	2400	1272	1399	1144	381	0	34,672
2	3957	4115	1582	2215	2849	4392	4708	3798	6924	6608	5025	8507	4392	5342	1899	4115	2849	949	316	70,585
3	6697	16,005	8505	5893	10,648	7433	3214	3750	3214	11,183	10,684	12,791	7969	6964	7969	5893	2678	4821	1607	131,185
4	9663	13,818	8503	10,725	5411	5411	7730	3092	8503	8503	10,725	8503	12,272	3092	10,049	5411	9276	3865	6597	141,846
5	11,272	7214	7214	4508	11,722	8115	3607	901	5410	6312	4508	6312	8115	901	10,821	2705	9919	5410	8115	111,809
6	15,496	2479	14,876	13,636	7438	3719	7438	3719	4958	1239	8677	8677	8677	8677	7438	4958	1239	6198	32,076	146,119
7	17,589	0	8442	2814	0	0	0	0	0	0	0	0	0	0	2814	0	2814	2814	16,885	36,583
8	21,613	0	6916	1729	0	0	1729	0	0	1729	0	0	0	0	0	0	0	0	5187	17,290
9	24,360	0	3897	0	0	0	0	0	0	0	0	0	0	0	0	0	0	0	0	3897
Total KE (erg/cm^2)		45,030	61,827	44,684	40,341	32,488	31,590	18,932	30,901	37,593	41,729	45,935	43,317	27,376	42,262	24,481	29,919	24,438	71,143	693,986
Total KE ($J\ m^{-2}\ mm^{-1}$)		33·9	36·4	35·2	33·8	32·7	33·7	31·6	33·3	33·2	34·0	33·2	34·3	33·3	34·8	33·6	34·1	37·1	38·4	34·6

There was a positive correlation ($r = 0{\cdot}70$) between the rainfall intensity and the average drop volume, and as mentioned above, 49 per cent of the increase in the rainfall intensity may be attributed to the increase in drop size.

3.2.3.4. Energy Load

The energy load distribution of the rainstorm in successive one-minute intervals is shown in Table 3.9, while the kinetic energy load as related to the drop size distribution for the rainstorm as a whole is shown in Table 3.6. The kinetic energy load ranged from 31·6 to 38·4 J m^{-2} mm^{-1} rainfall, while the average kinetic energy load for the whole rainstorm was 34·6 J m^{-2} mm^{-1} (Table 3.9). These values are much higher than those quoted by Elwell and Stocking (1973) for Rhodesia, where the average load for the season with 910 mm rainfall was of the order of 19 J m^{-2} mm^{-1}. This suggests that the rainfall in the study area in Rhodesia may be predominantly composed of drops less than 2 mm diameter, rather similar to the drop size distribution pattern found in England (Bean and Wells, 1953). On the other hand, the greater energy load of rainstorms at Samaru is due to a much greater proportion of large-volume drops, 59 per cent of all drops being more than 3 mm in diameter.

There was a positive correlation ($r = 0{\cdot}99$) between the amount of rainfall (R_a) and the kinetic energy load, and between the rainfall intensity (R_i) and the kinetic energy load. This suggests that the kinetic energy load of rainstorms may be estimated from either the rainfall intensity or rainfall amount. The regression equation relating kinetic energy (KE) in J m^{-2} to rainfall amount (R_a) was:

$$\mathrm{KE} = 38{\cdot}8R_a - 4{\cdot}7 \times 10^{-4}. \qquad (3.11)$$

The negative intercept of 4·7 J/m^{-2} suggests that a rainfall of 0·12 mm has zero kinetic energy load, which is a relatively small error.

It was found, however, that both the intensity and amount of rainfall were poorly correlated ($r = 0{\cdot}56$) with the energy load expressed in J $m^{-2\,mm-1}$ rainfall. On the other hand, the energy load of rainstorms calculated from the ratio R_v/D_n (i.e. average drop volume) and the corresponding terminal velocity of the average drop closely approximated the energy load values derived from the sum of products of the kinetic energy of drops in different classes and their corresponding drop numbers.

3.2.4. ENERGY LOAD, EROSIVENESS AND SEASONAL CHARACTERISTICS OF RAINSTORMS

3.2.4.1. Energy Load

The records of the 18 rainstorms taken in 1973 and 1974 were analysed and summarized in a similar manner to those in Table 3.6. The amount of rainfall and the kinetic energy load of all these 18 rainstorms are shown in Table 3.10. As expected from the previous analysis, there was a significant ($P = 0{\cdot}1$) positive correlation ($r = 0{\cdot}85$) between the amount of rainfall (R_a) and the average drop volume (D_v). For the range of rainfall between 6·60 and 60·64 mm, the regression equation was:

$$D_v = 1{\cdot}77 + 0{\cdot}93R_a. \tag{3.12}$$

The line did not pass through the origin, indicating the insensitivity of the instrument to register drops smaller than 1·5 mm in diameter.

Table 3.10. Amount of rainfall (R_a) and kinetic energy load (KE) of 18 rainstorms

R_a (mm)	KE ($J\ m^{-2}$)	KE ($J\ m^{-2}\ mm^{-1}$)
41·64	1599	38·40
20·07	693	34·58
16·10	584	36·29
751	751	36·46
60·64	2394	39·49
6·60	149	22·59
37·52	1423	37·94
28·10	1024	36·48
13·21	390	29·57
6·60	143	21·81
9·05	264	29·24
15·20	482	31·72
16·51	630	38·18
12·19	356	29·24
8·89	221	24·90
6·60	203	30·84
33·02	1264	38·29
7·62	180	23·68

$KE = 41{\cdot}4R_a - 120{\cdot}0\ J\ m^{-2}$, $r = 0{\cdot}999$

Using equation (3.12), it is possible to calculate the energy load of individual storms, taking into account 72 per cent of the variation and without measuring the rainfall intensity. For example, the average drop volume of rainfall of 20 mm calculated from equation (3.12) would be 20·4 mm^3. The average number of drops per cm^2 would be 2000/20·4 = 98·2. The terminal velocity of a 20·4 mm^3 drop is 8·28 m/sec (taken from prepared tables). Then, the kinetic energy of the average drop is 6983 erg and of the rainstorm, 6983 × 98·2 = 685,000 erg/cm^2 or 685 J/m^2 or 34·3 $J\ m^{-2}\ mm^{-1}$.

Equation (3.12) may be transformed to give kinetic energy load (KE) of rainstorms in $J\ m^{-2}$ from rainfall amount (R_a) by substituting the drop volume values with the corresponding kinetic energy values as follows:

$$KE = 43{\cdot}0R_a - 155{\cdot}0 \tag{3.13}$$

The kinetic energy load values of the 18 rainstorms (Table 3.10) obtained from an analysis of drop size and number was highly correlated ($r = 0{\cdot}99$) with the corresponding rainfall values in mm. The regression is given in equation (3.14).

$$\mathrm{KE} = 41{\cdot}4R_a - 120{\cdot}0 \qquad (3.14)$$

The regression coefficients of both the equations (3.13) and (3.14) are similar. Although the negative intercepts of 155 and 120 J m^{-2} in these equations suggest that rainstorms with total rainfall in the range of 2·9 and 3·6 mm have zero kinetic energy, this is acceptable in runoff and erosion studies.

3.2.4.2. Seasonal Distribution of Energy Load

The seasonal distribution pattern of rainfall at Samaru has been described by Kowal and Knabe (1972). Using average rainfall values from the last 15 years' rainfall records, the magnitude of the kinetic energy load (J/m^2) of rainstorms for each month of the wet season at Samaru was estimated from equation (3.14). This was done by calculating the energy load of each rainstorm greater than 4 mm and summing the energy load values for each month. The seasonal distribution of the kinetic energy load is shown in Figure 3.2.

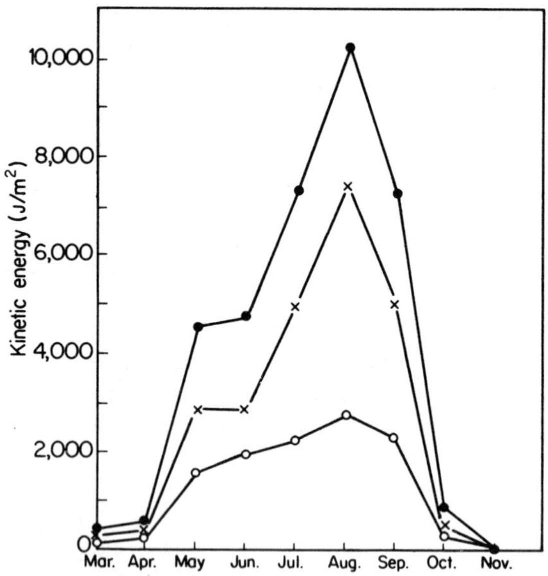

Figure 3.2. Kinetic energy load of rainstorms at Samaru during the wet season. Energy load of total rainfall (●) of 1091 mm = 35,975 J/m^2. Energy load of erosive rainfall (×) of 635 mm = 24,180 J/m^2. Energy load of non-erosive rainfall (○) of 456 mm = 11,790 J/m^2

The total kinetic energy load of 1091 mm (42·95 in.) total rainfall at Samaru was about 36,000 J/m^2. This is twice as large as the amount estimated by Elwell and Stocking (1973) for the Henderson Research Station, Rhodesia. The seasonal distribution of the kinetic energy load is similar to the seasonal distribution of the amount of rainfall, culminating in a peak in August at 10,300 J/m^2 when the precipitation is 302 mm (11·9 in.).

The average kinetic energy of rainstorms greater than 4 mm is about 33 J m^{-2} mm^{-1}. This suggests that the predominant drop size is about 3·3 mm diameter.

3.2.4.3. Erosiveness of Rainstorms

Kowal (1970a, b), studying surface runoff and soil erosion from cultivated areas of gentle and long slopes at Samaru, found that out of an average of 85 rainy days in the wet season, 32 rainstorms (38 per cent) produced runoff. The water from the remaining 53 rainstorms was wholly accepted by the soil. All storms

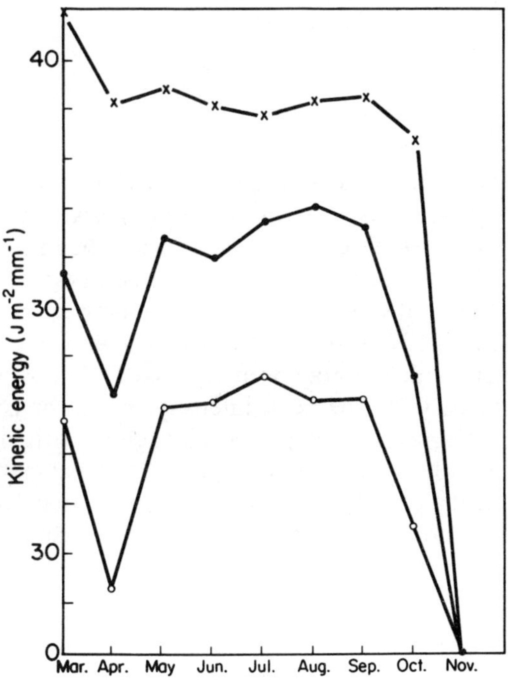

Figure 3.3. Wet season distribution of kinetic energy of rainstorms (monthly average) at Samaru. Annual average energy load for all (●), erosive (×) and non-erosive (○) rainstorms equal to 33·0, 38·1 and 25·9 J m^{-2} mm^{-1}, respectively

exceeding 20 mm of rainfall resulted in runoff, the degree of runoff depending on the rainfall duration and intensity.

The study by Kowal (1970a, b) therefore allows a broad generalization to be made about the erosiveness of rainstorms on cultivated land at Samaru, i.e. surface runoff and soil erosion occur when rainfall is greater than 20 mm. Using this distinction, the seasonal distribution of the energy load of erosive and non-erosive rainstorms was calculated using 15 years rainfall records (Figures 3.2 and 3.3). On this basis, about 58 per cent of the annual rainfall (635 mm) was erosive and had a total energy load of about 24,200 J/m^2 (Figure 3.2), corresponding to an average of 38 $J\,m^{-2}\,mm^{-1}$ with monthly average values showing minor variation from the annual average value (Figure 3.3). This suggests that the predominant drop size of the erosive rainfall is about 4 mm diameter.

Non-erosive rainfall was about 42 per cent of the total annual rainfall and contained 33 per cent of the total annual energy load. The average energy load of the non-erosive rainfall was 25·9 $J\,m^{-2}\,mm^{-1}$ (Figure 3.3) which suggests a predominant drop size of 2·5 mm diameter.

From the seasonal distribution of the erosive rainstorms (Figure 3.3), it is evident that the risk of soil erosion varies little throughout the wet season. However, the frequency of occurrence of the erosive rains is much higher in the months of July, August and September.

3.2.4.4. Rainfall Intensity

Hudson (1963) found that the rainfall intensity of about 25 mm/hr was a threshold value, higher intensities than this being erosive. In the study by Elwell and Stocking (1973) an empirical formula, relating the rainfall intensity to the kinetic energy load, was used to deduce the erosiveness of rainfall. Previous studies at Samaru have not shown a direct relation between rainfall intensity and soil erosion. Records of 157 rainstorms for which peak intensity was recorded by a 68·6 cm diameter Jardi recorder were analysed. A positive correlation ($r = 0{\cdot}78$) was found between the peak intensity in mm/hr and the amount of rainfall in mm. The regression line passed through the origin with a regression coefficient of 0·23 mm/hr increase in peak rainfall intensity per 1 mm increase in rainfall. As mentioned already, Kowal (1970a, b) found that runoff occurred only when rainfall was greater than 20 mm. This would correspond to peak intensity values greater than 88 mm/hr.

3.2.5. CONCLUSIONS

The study was initiated to provide information on the erosive nature of rainstorms in the tropics and to identify easily measurable rainfall parameters for assessing the energy load of tropical rainstorms. The instrument and analyses employed in this study appear to be of sufficient accuracy for studies of runoff and soil erosion. The instrument, however, is not satisfactory for an accurate assessment of the drop size distribution for non-tropical rainstorms.

The relationships between the volume of rainfall and the drop volume, and between the volume of rainfall and the mean number of drops, held true for rainstorms ranging from 4 to 60 mm. It is not known whether these relationships would hold true for rainstorms greater than 60 mm. These relationships do not apply to drizzles or intermittent traces of rainfall.

The relatively greater kinetic energy load of rainstorms at Samaru is due to a greater proportion of larger diameter drops. About 60 per cent of all the drops in a typical rainstorm were greater than 3 mm diameter. In part, this is probably due to the higher rainfall intensities, which would lead to a greater concentration of drops in the air space and a greater occurrence of collisions and fusion between drops. The other reason for larger drop size might be that most of the precipitation at Samaru originates from cumulus or cumulonimbus clouds of considerable vertical depths and high condensation intensities. Under such conditions, the chances for drops to grow are higher.

ACKNOWLEDGMENTS

The work reported here forms part of a research programme approved by the Board of Governors of the Institute for Agricultural Research, Ahmadu Bello University, Samaru, Nigeria.

The authors are most grateful to Dr M. Dagg, Director of the Institute, for kindly critically reading the manuscript and for his useful suggestions.

SUMMARY

Using a specially designed instrument to give a continuous record of the number and size of raindrops on a time-scale, the instantaneous intensity and energy load of tropical rains were assessed. Both the intensity and drop size of rains in the West African savannah are much greater than in temperate climates. The high intensity of tropical storms is due to the larger size and greater number of raindrops falling per unit time. Because the predominant drop size of tropical rains is large, the energy load of storms is high. A high correlation between the average drop volume of rainfall and the amount of rainfall allows for a reliable and convenient estimate of the energy load of storms from the amount of rainfall. The seasonal distribution of the energy load of rains in the West African savannah is similar to the annual distribution of rains. About 58 per cent of the total annual rainfall is erosive, containing two-thirds of the total annual energy load of the rains.

NOTES

1. This article originally appeared in *Tropical Agriculture*, 1976, **53** (3), 185–198 and is reproduced by permission of *Tropical Agriculture*.

REFERENCES

Bean, A. G., and D. A. Wells, 1953, Soil capping by water drops, *Nat. Inst. Agric. Eng. Rep.*

Ellison, W. D., 1944, Studies of raindrop erosion, *Agric. Eng.*, **25**, 131–136.

Ellison, W. D., 1952, Raindrop energy and soil erosion, *Emp. J. Exp. Agric.*, **20**, 81–86.

Elwell, H. A., and M. A. Stocking, 1973, Rainfall parameters for soil loss estimation in a sub-tropical climate, *J. Agric. Eng. Res.*, **18**, 169–177.

Hudson, N. W., 1963, Rain drop size distribution in high intensity storms, *Rhod. J. Agric. Res.*, **1**, 6–11.

Hudson, N. W., 1971, *Soil Conservation*, Batsford, London.

Kowal, J., 1970a, The hydrology of a small catchment basin at Samaru, Nigeria. III. Assessment of surface run-off under varied land management and vegetation cover, *Niger. Agric. J.*, **7**, 120–133.

Kowal, J., 1970b, The hydrology of a small catchment basin at Samaru, Nigeria. IV. Assessment of soil erosion under varied land management and vegetation cover, *Niger. Agric. J.*, **7**, 134–147.

Kowal, J., W. Kijewski and A. H. Kassam, 1973, A simple device for analysing the energy and intensity of rainstorms, *Agric. Meteorol.*, **12**, 271–280.

Kowal, J., and D. Knabe, 1972, *An Agro-climatological Atlas of the Northern States of Nigeria*, Ahmadu Bello University Press, Samaru.

Kowal, J., and K. S. Stockinger, 1973, The usefulness of ridge cultivation in Nigerian agriculture. *Soil Water Conserv. J.*, **28**, 136–137.

3.3

Susceptibilities of Certain Tropical Soils to Erosion by Water

S. A. El-Swaify

3.3.1. INTRODUCTION

Quantitative assessment of soil erosion hazards requires full knowledge of many factors related to soil properties, land conditions, and rainfall characteristics. The universal Soil Loss Equation (Wischmeier and Smith, 1965) expresses the relationship between these factors quantitatively in the form:

$$A = RKLSCP \tag{3.15}$$

in which A is the mass of soil lost per unit area, R the rainfall factor, K the soil erodibility factor, L the length of slope factor, S the slope gradient factor, C the cropping management factor, and P is the factor related to erosion control practices. Values for each of these factors have been collected, but mainly for the mainland United States. The humid tropics differ from other regions as already shown in rainfall characteristics (consequently requiring different R indices), and may be expected to differ on soil characteristics (with different K values), and prevailing vegetation or crops (with different C factors). Whether slope gradients and lengths (whose effects are sometimes combined in a general LS factor) and erosion control measures (expressed in the P factor) for these soils follow trends determined for temperate zone soils is yet to be established. The purpose of this chapter is to present some of the information now available on the susceptibilities of selected soils of the humid tropics to erosion by water.

3.3.2. THE SOILS

Detailed data presented here were obtained for selected soils in the State of Hawaii. The soils were from 10 families within the orders Aridisols, Inceptisols, Oxisols, Ultisols, and Vertisols. The series were Kawaihae very fine sandy loam (medial isohyperthermic Ustollic Camborthid), Hilo silty clay loam (thixotropic isohyperthermic Typic Hydrandept), Kukaiau silty clay loam (thixotropic isothermic Hydric Dystrandept), Naalehu extremely stony silty clay loam (medial isohyperthermic Typic Eutrandept), Pakini very fine sandy loam (medial isohyperthermic Entic Eutrandept), Waipahu silty clay (very fine, kaolinitic, isohyperthermic Vertic Ustropept), Molokai silty clay loam (clayey, kaolinitic, isohyperthermic Typic Torrox), Wahiawa silty clay (clayey, kaolinitic, isohyperthermic Tropeptic Eutrustox), Waikane silty clay (clayey, kaolinitic, isohyperthermic Humoxic Tropohumult), and Lualualei clay (very fine, montmorillonitic, isohyperthermic Typic Chromustert). The first five are volcanic ash soils from the island of Hawaii and the remaining ones are from the island of Oahu. Further information on these soils may be found in the Soil Survey of the State of Hawaii (Foote *et al.*, 1972; Sato *et al.*, 1972).

3.3.3. PROCEDURES

A rainfall simulator similar to that described by Meyer and McCune (1958) was used in these studies. Bare plots were selected on sites with slope gradients of 3 to 18 per cent and bordered to enclose areas of 3·66 m width and either 10·67 or 22·86 m length. The plots were uniformly prepared for all sites to simulate fallow conditions before two consecutive storms with intensities of approximately 63 mm/hr were applied. The first storm, called a 'dry run', was applied at the prevailing soil water content at the time and the second, called a 'wet run', was applied 16–24 hr later at soil water contents between saturation and field capacity. The wet run, applied in these studies to simulate the cyclonic conditions typical of many tropical rainstorms, is believed to be important for realistic simulation of soil losses in the humid tropics. It has recently been shown (Barnett and Dooley, 1972; Young and Burwell, 1972) that soil losses induced by this rainfall simulator are comparable to those induced by natural rainfall in two widely separated regions (the southeast and midwest) of the United States.

3.3.4. RESULTS AND DISCUSSION

Actual soil loss data for each study site and calculated erodibilities for all the above soils have been published elsewhere (Dangler, El-Swaify and Barnett, 1975; Dangler, El-Swaify, Ahuja and Barnett, 1975). Data presented here have been generated by use of those erodibility factors for a variety of conditions selected for application of equation (3.15).

3.3.4.1. Comparative Erosion Susceptibilities of Soils

To illustrate differences among soils, equation (3.15) may be applied to calculate predicted soil losses from one hectare of fallow land with a 9 per cent gradient and an average slope length 22·1 m and which receives no erosion control practices. In this manner, a value of 1·0 may be assigned to the C, P, and combined (LS) factors (Wischmeier and Smith, 1965). The losses were calculated for a rainstorm with uniform intensity of 63·5 mm/hr and duration of 1 hr. When all the above factors are substituted in equation (3.15), soil losses under this hypothetical storm may be calculated as follows using known erodibility factors:

$$A = 146{\cdot}8 \times K \text{ tonnes/ha.} \tag{3.16}$$

Table 3.11 shows that two different sets of figures may be obtained for each soil. The first, obtained from what is described above as a dry run, is probably more applicable during the summer months in Hawaii. The second, called a wet run, is more applicable to the winter months. Distinctions between the two values were not necessary for all soils and may be explained, at least in part, by the faster occurrence of runoff during the wet runs. Furthermore, it appeared that poorly structured soils (e.g. Kawaihae, Pakini) or those that exist in a continually wet condition (Hilo) are almost equally susceptible to water erosion regardless of the prevailing soil moisture regime.

Table 3.11. Predicted soil losses from selected Hawaii soils subjected to a 1 hr storm with 63·5 mm/hr intensity*

Soil	Classification subgroup	Soil loss (tonnes/ha) Dry season	Wet season
Kawaihae	Ustollic Camborthids	52·9	54·3
Hilo	Typic Hydrandepts	10·7	10·3
Kukaiau	Hydric Dystrandepts	17·6	32·3
Naalehu	Typic Eutrandepts	23·5	38·2
Pakini	Entic Eutrandepts	88·1	74·9
Waipahu	Vertic Ustropepts	4·40	60·2
Molokai	Typic Torrox	20·6	36·0
Wahiawa	Tropeptic Eutrustox	13·2	29·4
Waikane	Humoxic Tropohumults	1·47	10·3
Lualualei	Typic Chromusterts	38·2	45·5

* See text for assumed values of factors in the universal soil loss equation.

The range of losses encountered in the above soils as listed in Table 3.11 is a clear indication of the wide differences to be expected among susceptibilities of various tropical soils. Even within a single soil order, the range of variation can be as wide as it is between different orders. For instance, the order Inceptisols includes both the most erodible soil investigated (Pakini) and the second least erodible (Waipahu). As a whole, the relative erodibilities of the soil subgroups in this study to rainstorms occurring in the dry season appear to be Humoxic Tropohumults < Vertic Ustropepts < Typic Hydrandepts < Tropeptic Eutrustox < Hydric Dystrandepts < Typic Torrox < Typic Eutrandepts ≪ Typic Chromusterts < Ustollic Camborthids < Entic Eutrandepts. This order changes appreciably when susceptibilities to erosion during the wet season are considered, becoming Humoxic Tropohumults = Typic Hydrandepts ≪ Tropeptic Eutrustox < Hydric Dystrandepts < Typic Torrox = Typic Eutrandepts < Typic

Chromusterts < Ustollic Camborthids < Vertic Ustropepts < Entic Eutrandepts. Therefore, depending on the probability of a 60 mm rainstorm occurring on the defined soil conditions above, it would appear that only 3–4 of the subgroups meet the annual tolerance criterion (12 tonnes/ha) under dry conditions and only 2 under wet conditions (Wischmeier and Smith, 1965).

When values in Table 3.11 are compared with those available in the literature for other soils, it is noted that the range of variations in susceptibilities of tropical soils is as wide as it is for temperate zone soils. Using descriptive classes as adopted by Dangler, El-Swaify, Ahuja and Barnett (1975), erosion hazards for soils undergoing losses of 0–14·6, 14·7–29·2, 29·4–46·8, 47·0–63·0, 63·1–80·6, and >80·7 tonnes/ha under the storm defined here may be described as very low, low, moderate, moderately high, high, and very high respectively. All these classes were encountered in the studies reported here. Wischmeier and Smith (1965) reported erodibility factors for a variety of soils on the US mainland which would result in predicted losses ranging from 4·40 to 101 tonnes/ha, thus also covering the above classes. The few available data on erodibilities of other tropical soils indicate some agreement with values reported here for related soils from Hawaii. Barnett *et al.* (1971) reported erodibilities for two Inceptisols (Vertic Eutropepts and Typic Dystropepts) and an Ultisol (Typic Tropohumult) from Puerto Rico. Their values, for the conditions defined here, would result in predicted soil losses of 0·587, 2·50, and 16·6 tonnes/ha for the above three soils, respectively. Little or no emphasis was placed by previous workers on the effect of soil water regime on erodibility. This may have been due to the very low statistical significance of antecedent water contents reported for certain temperate soils (Wischmeier and Mannering, 1969). Clearly, this was not true for most Hawaiian soils.

Soil losses from sites subjected to normal agricultural operations were found to be less than those subjected to disturbance such as occurs during construction activities. This was particularly noted under the dry moisture regime, i.e. for losses incurred during the first storm. Losses from the disturbed Typic Torrox (Molokai) soil were 60·2 and 45·5 tonnes/ha for the dry and wet regimes, respectively, indicating that losses have trebled for the dry regime and increased slightly for the wet regime over corresponding values in Table 3.11.

3.3.4.2. Erosion Losses in Relation to Soil Properties

Aside from land-related factors as incorporated in equation (3.15), the erodibility of a soil is considered an inherent property which depends on numerous hydrologic, textural, and structural parameters. Achieving an understanding of how each of these parameters affects erodibility is valuable because it allows predictions of soil susceptibility to erosion with less effort, expense, and time than are required for experimental assessment either under natural rainfall or, as reported here, under simulated rainfall. It is for this reason that many workers approached this problem by attempting to establish erodibility indices, rather than erodibility factors, for direct use in quantitative soil loss predictions (Yamamoto and Anderson, 1973). Nevertheless, equations have been published by some authors for prediction of soil erodibility factors from selected basic soil parameters or from combinations thereof (e.g. Barnett and Rogers, 1966; Wischmeier and Mannering, 1969). It has been generally found that a very large number of such parameters is necessary to achieve a reasonably full account of soil erodibility. In some instances, erodibility approximations were possible by use of only a limited number of parameters (Wischmeier *et al.*, 1971). No such predictive equations have been developed for tropical soils, nor have the above available predictive methods been tested on such soils. Table 3.12 shows a preliminary test of the nomograph developed by Wischmeier *et al.* (1971) for estimating losses from the soils in this study under the same storm and conditions previously defined for Table 3.11. Serious discrepancies are apparent between the data in both tables, particularly for soils with very low or high susceptibilities to erosion. This is not surprising in view of the fact that highly weathered tropical soils have different mineralogical and structural properties that would be expected to affect erodibility. Of particular interest is the fact that neither the nomograph nor the original predictive equations (Wischmeier and Mannering, 1969) include a measure of the contributions of amorphous or oxidic constituents to structural stabilities, and, thus, susceptibility to erosion (El-Swaify, 1970; El-Swaify and Emerson, in press; Jones and Uehara, 1973; Twenboah *et al.*, 1969).

Table 3.12. Soil losses from selected Hawaii soils* by use of predicted *K*-factors†

Soil	Soil loss (tonnes/ha) based on		
	Apparent texture	H_2O dispersion	NaOH dispersion
Kawaihae	36·7	32·3	35·2
Hilo	10·3	4·40	4·40
Kukaiau	22·0	13·2	8·81
Naalehu	25·0	27·9	16·2
Pakini	11·7	11·7	7·34
Waipahu	20·6	14·7	14·7
Molokai	22·0	13·2	11·7
Wahiawa	10·3	27·9	7·34
Waikane	22·0	39·6	7·34
Lualualei	19·1	19·1	19·1

* For the same conditions as specified for calculations from experimental data.
† Using the nomograph proposed by Wischmeier *et al.* (1971).

It is concluded that susceptibilities of tropical soils to erosion by water depend on as yet undetermined properties or interrelations of properties. Establishing such properties or interrelations will be necessary for successful predictions of these susceptibilities by simplified methods.

SUMMARY

Soil erodibility factors, experimentally obtained by rainfall simulation, were used to evaluate the relative susceptibilities of selected Hawaii Aridisols, Inceptisols, Oxisols, Ultisols, and Vertisols to erosion by water. For this evaluation, soil losses from a defined area of 1 ha due to a hypothetical 1 hr storm with a uniform intensity of 63·5 mm/hr were predicted by the universal soil loss equation. It was found that susceptibilities of the soils to water erosion covered a wide range of values from very low (less than 14·6 tonnes/ha) to very high (more than 80·7 tonnes/ha). For most soils, it was necessary to make a clear distinction between susceptibilities to erosion depending on the prevailing soil moisture regime. Nevertheless, in general it appeared that the Humoxic Tropohumult was the least, and the Entic Eutrandept the most, susceptible.

Quantitative predictions of soil losses could not be adequately made by use of the parameters and relationships successfully used for temperate soils. Therefore, additional parameters affecting erodibility need to be defined and more applicable predictive equations developed for soils of the humid tropics.

NOTES

1. This chapter is a contribution from the Department of Agronomy and Soil Science, Hawaii Agricultural Experiment Station Journal Series No. 1892 and Hawaii Institute of Geophysics Contribution No. 677. The work was supported in part by Cooperative Agreement No. 12-14-5001-40 with the Agricultural Research Service, USDA.

REFERENCES

Barnett, A. P., J. R. Carreker, F. Abruna and A. E. Dooley, 1971, *Trans. ASAE*, **14** (3), 496–499.

Barnett, A. P., and A. E. Dooley, 1972, *Trans. ASAE*, **15** (6), 1112–1114.

Barnett, A. P., and J. S. Rogers, 1966, *Trans. ASAE*, **9** (1), 123–128.

Dangler, E. W., S. A. El-Swaify, L. R. Ahuja and A. P. Barnett, 1975, ARS-W-Series, Western Region ARS, USDA.

Dangler, E. W., S. A. El-Swaify and A. P. Barnett, 1975, *Hawaii Agric. Exp. Sta. Research Bulletin* 181.

El-Swaify, S. A., 1970, *Soil Sci.*, **10** (3), 197–202.

El-Swaify, S. A., and W. W. Emerson, 1975, *Soil Sci. Soc. Amer. Proc.*, **39**, 1056–1063.

Foote, D. E., E. L. Hill, S. Nakamura and F. Stephens, 1972, Soil survey of islands of Kauai, Oahu, Maui, Molokai, and Lanai, State of Hawaii, USDA, SCS, in cooperation with the University of Hawaii Agricultural Experiment Station.

Hudson, N., 1971, *Soil Conservation*, Cornell University Press, Ithaca, New York.

Jones, R. C., and G. Uehara, 1973, *Soil Sci. Soc. Amer. Proc.*, **37**, 792–798.

Meyer, L. D., and D. L. McCune, 1958, *Agricultural Engineering*, **39** (10), 644–648.

Sato, H., W. Ikeda, R. Paeth, R. Smythe and Minoru Takehiro, Jr., 1972, Soil survey of island of Hawaii, State of Hawaii, USDA, SCS, in cooperation with the University of Hawaii Agricultural Experiment Station.

Twenboah, C. K., J. W. Kijne and D. J. Greenland, 1969, *Aust. J. Soil Res.*, **7**, 325–331.
Wischmeier, W. H., C. B. Johnson and V. B. Cross, 1971, *Jour. Soil Water Cons.*, 189–193.
Wischmeier, W. H., and J. V. Mannering, 1969, *Soil Sci. Soc. Amer. Proc.*, **33**, 131–137.
Wischmeier, W. H., and D. D. Smith, 1965, Agric. Handbook No. 282, US Govt. Printing Office, Washington, D.C.
Yamamoto, T., and H. W. Anderson, 1973, *Water Resour. Res.*, **9** (2), 336–345.
Young, R. A., and R. E. Burwell, 1972, *Soil Sci. Soc. Amer. Proc.*, **36**, 827–830.

Part 4

Management for Erosion Control

4.1

Soil-Conserving Versus Soil-Degrading Crops and Soil Management for Erosion Control

R. LAL

As discussed in previous chapters, erodibility and erosivity are inherent properties of soil and climate respectively, and little can be done to alter them. Erosion control therefore depends on judicious soil and crop management, i.e. the *C* and *P* factors of the Universal Soil Loss Equation.

4.1.1. SOIL MANAGEMENT PRACTICES

Soil or land management practices are based on two broad principles:

(i) those practices which help maintain soil infiltration rates at sufficiently high levels to reduce runoff to a negligible amount; and

(ii) practices which help safe disposal of runoff water from the field, should rainfall exceed the infiltration capacity of the soil.

The choice of any particular technique depends on various factors. Usually a combination of both high infiltration rates and measures to dispose of runoff safely will be needed for adequate erosion control.

Cultural practices which help maintain a high soil infiltration rate are essentially based on farming techniques which maintain a mulch or live vegetation on the soil, such as stubble mulching and no-tillage or minimum tillage, and use of cover crops. The safe disposal of runoff water may involve physical manipulation of soil, including land-shaping, construction of contour bunds (Plate 5), terraces (Plate 6), waterways, ridges or tied ridges (Hudson, 1957).

The infiltration capacity and the potential infiltration rate of a majority of the soils in the humid tropics are high (Wilkinson and Aina, 1976). Equilibration rates of 20 to 25 cm/hr are not uncommon. Slaking and subsequent crust formation at the soil surface as a result of direct raindrop impact decreases both the infiltration capacity and the rate, and so results in accelerated runoff and erosion (Table 2.1, p. 19.)

Hudson (1957) and recently Elwell and Stocking (1976) have demonstrated for tropical conditions that protection from raindrop impact, as provided by vegetal cover, maintains the soil infiltration rate. Hudson (1957) reported for a ferrallitic sandy-clay loam soil on 4·5 per cent slope that soil loss from the unprotected soil was 127 times that from a plot covered with mosquito-gauge nylon net, and the runoff was 13 times greater. A plot with grass cover gave results similar to the one with mosquito-gauge net cover (Table 4.1).

Table 4.1. Soil loss and runoff from protected and unprotected soil (after Hudson, 1957)

Mean annual values	Bare ground	Soil protected with mosquito-gauge net
Soil loss (tonnes/ha/yr)	127	1
Runoff (mm)	271	21
Runoff (% rainfall)	32·2	2·5

This implies that some form or other of cover on the soil surface is important to protect against runoff and soil loss. Lal (1976b) reported the runoff and soil loss from bare fallow plots and plots covered with 6 tonnes/ha of straw mulch (Table 4.2) for an Alfisol near Ibadan in Nigeria. The runoff from the bare plots was 16 times more than that from the mulched plots. The mean annual soil loss from the mulched plots, however, was 730 times less than that from the bare

Table 4.2. Effect of ground cover on runoff and soil loss for 10% slope (after Lal, 1976b)

Mean annual values	Bare ground	Mulched
Soil loss (tonnes/ha/yr)	232·6	0·2
Runoff (mm)	504·1	29·3
Runoff (% rainfall)	42·1	2·4

Table 4.3. Relative efficiency of soil surface treatments in preventing runoff (modified from Lawes, 1966)

Soil surface treatment	Runoff (%)	Maximum infiltration rate (mm/hr[1])
Bare soil		
Undisturbed	69	12·7
Hoed at fortnightly intervals	51	22·9
Hoed after every storm	43	30·5
Mulches		
Dead grass	10	Greater than 127
Groundnut shells	11	Greater than 127
Sorghum stalks	2	Greater than 127

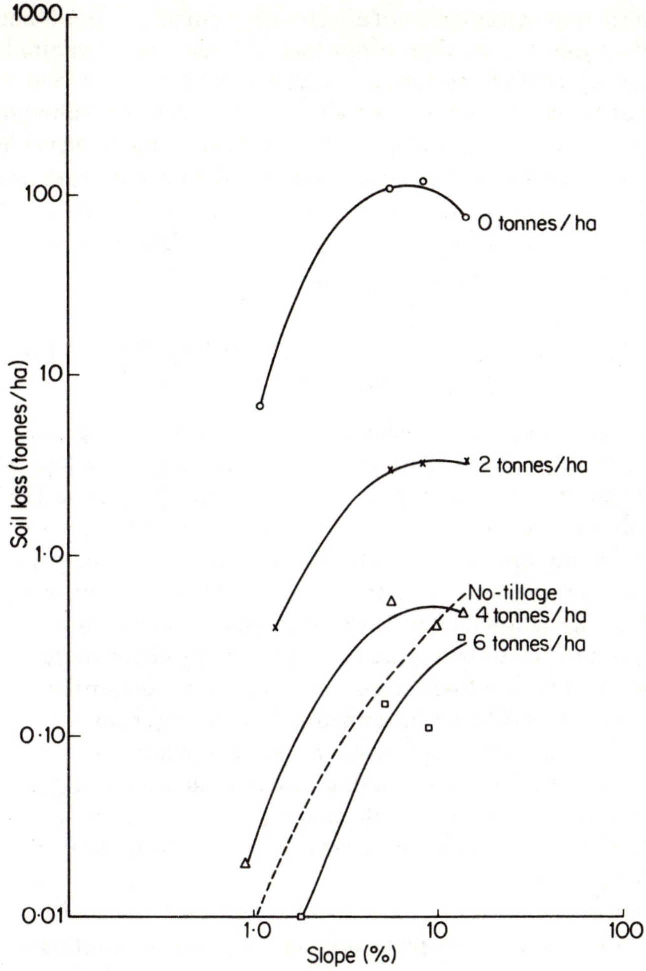

Figure 4.1. Effect of mulch rate on soil loss under different slopes

ground (see also Figure 4.1). That mulch cover maintains the infiltration rate of soils in tropical regions has also been demonstrated by Lawes (1966) (Table 4.3). These results confirm much earlier work from other parts of the world. For the tropical soils, therefore, management practices based on maintaining soil infiltration rate can be of great significance in controlling runoff and soil loss.

The practice of tied ridging (Plate 7) has been reported to be important in increasing the time for the water to penetrate into the soil (Macartney *et al.*, 1971; Prentice, 1946). However, contour ridges made on sandy loam soils may or may not be successful. If one ridge breaks, it can result in more soil loss than flat cultivation (Plate 8). Kowal (1970a, b) reported from the savannah region of

Northern Nigeria that ridges were not effective in controlling runoff and soil loss. While flat cultivation under arable crops lost 20·3 per cent of rainfall as runoff carrying 4·0 tonnes/ha/yr of soil, ridges down slope had 28 per cent runoff and about 20 tonnes/ha/yr of soil loss, and alternate tied ridges resulted in 13·2 per cent runoff and 5·7 tonnes/ha/yr of soil loss. In Upper Volta, however, tied ridges lead to only 0·9 per cent runoff as compared with 6·3 per cent with open graded ridges and 12·2 per cent in the case of flat cultivation (Fournier, 1967). The efficacy of the technique will obviously depend on details of erodibility and erosivity, and on cropping methods.

4.1.2. CROP MANAGEMENT: SOIL-DEPLETING VERSUS SOIL-CONSERVING CROPS

Crops and management practices which help produce an early ground cover are certainly more useful in controlling runoff and erosion than those which take longer for full canopy cover to develop. For instance, Aina *et al.* (1976) found that the number of days required for 50 per cent canopy cover to form were approximately 38, 46, and 63, respectively, for cultivars of soybeans (*Glycine max*), pigeonpeas (*Cajanus cajan*), and cassava (*Manihot esculenta*) grown at Ibadan, Nigeria. The runoff and erosion losses which occurred under these crops were proportional to the time required for full canopy cover to develop. Crops with quick growth may be termed soil-conserving crops, as compared with those which take longer and may hence be termed soil-depleting crops, although it must be noted that differences will occur between cultivars of any one crop, depending on characteristics such as branching habit. Practices such as mixed cropping also affect ground cover. While it took sole cassava 63 days for 50 per cent ground cover to develop, mixed cropping of maize with cassava took only 51 days (Aina *et al.*, 1976). Soil erosion and runoff losses were proportionately less from mixed as compared with sole cropping systems (Table 4.4). Other cultural practices which affect erosion control are plant population, time of planting and fertility level (Hudson, 1957).

However, more important than the growth habit of a given crop is the soil management practice. A soil-depleting crop grown with proper soil-conserving techniques such as mulching could give rise to less runoff and soil loss than a soil-conserving crop grown with erosion-promoting practices. For instance, runoff

Table 4.4. Comparison of soil loss and runoff with monoculture (cassava) and mixed cropping of cassava with maize (Alfisol, Ibadan, Nigeria) (after Aina *et al.*, 1976)

Slope (%)	Soil loss (tonnes/ha/annum)		Runoff (%)	
	Monoculture	Mixed cropping	Monoculture	Mixed cropping
1	2·7	2·5	18	14
5	87·4	49·9	43	33
10	125·1	85·5	20	18
15	221·1	137·3	30	19

Plate 6. Terrace systems in the Philippines (courtesy Philippine Tourist and Travel Association)

[*Facing page* 84

Plate 7. The system of tied ridging can be successful on gentle slopes and for stable soils, although ponding of water behind the ties, as seen here, can be disadvantageous to crop growth

Plate 8. Soil erosion can be serious if a ridge system breaks on a sandy soil

Plate 9. Soil erosion under maize, a soil-degrading crop, grown with conventional ploughing and harrowing

Plate 10. Soybeans in the US grown through the maize residue from the previous crop

Plate 11. Use of crop residues as a soil-conserving technique in the US (courtesy Dr C. Larsun)

Plate 12. Increasing land prices make farmers reluctant to leave large areas of land which control systems require for water disposal

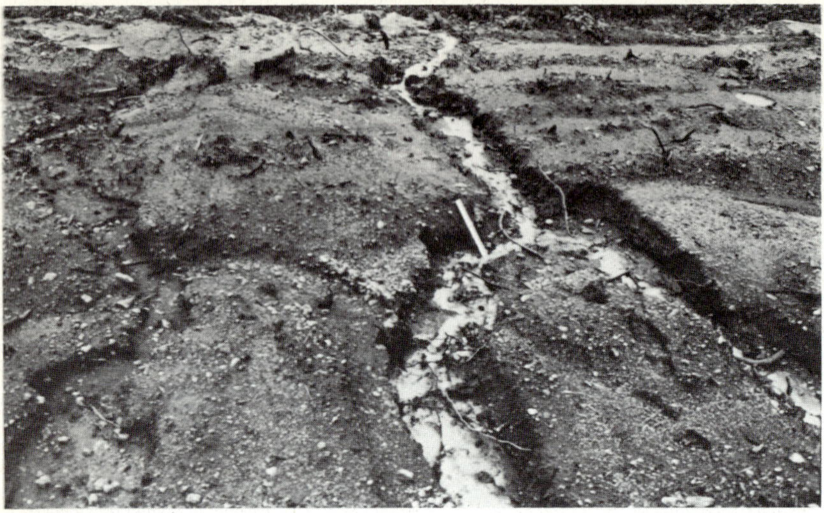

Plate 13. Accelerated soil erosion can occur on ploughed bare soil even on gentle slopes, but surface mulch (Plates 14, 15, 16) gives almost complete protection

Plate 14. Cowpeas (*Vigna unguiculata*) grown in the no-till system on an Alfisol near Ibadan, Nigeria

Plate 15. Maize grown in the no-till system, through a mulch of residues of the previous crop

Plate 16. Soybeans (*Glycine max.*) grown at Ibadan in the no-till system, with a mulch of residues from the previous crop

and soil loss from maize (*Zea mais*), a soil-depleting crop, was less than from cowpea (*Vigna unguiculata*) when the former was grown with no-till techniques and the latter with conventional methods of ploughing and harrowing (Table 4.5).

Table 4.5. Effect of interaction between nature of crop and soil management techniques on runoff and soil loss (Alfisol, Ibadan, Nigeria) (Lal, 1976b)

	Runoff (% of rainfall*)		Soil loss (tonnes/ha)	
Slope (%)	Maize (No-till)	Cowpeas (Conventional till)	Maize (No-till)	Cowpeas (Conventional till)
1	1·2	15·8	0·91	0·24
5	1·8	31·1	0·03	0·65
10	3·1	40·0	0·00	1·71
15	3·5	17·2	0·10	1·22

* Rainfall during cropping period, 781 mm.

The soil management practices, therefore, are more crucial to runoff and erosion control than selection of suitable crops. An erosion-promoting crop such as maize (Plate 9) can be grown without causing serious erosion provided that a soil management technique which helps maintain the infiltration rate of the soil and enables a surface mulch to be maintained is used (Plate 10).

REFERENCES

Aina, P. O., R. Lal and G. S. Taylor, 1976, Soil and crop management in relation to soil erosion in the rainforest region of Western Nigeria, Symposium Proceedings of the National Soil Erosion Conference 25–26 May, 1976, Lafayette, Indiana.

Elwell, H. A., and M. A. Stocking, 1976, Vegetal cover to estimate soil erosion hazard in Rhodesia, *Geoderma*, **15**, 61–70.

Fournier, F., 1967, Research on soil erosion and soil conservation in Africa, *Afr. Soils.*, **12**, 52–96.

Hudson, N. W., 1957, Erosion control research—Progress reports on experiments at Henderson Research Station 1953–56, *Rhod. Agric. J.*, **54**, 297–323.

Kowal, J., 1970a, The hydrology of a small catchment basin at Samaru, Nigeria. III. Assessment of surface runoff under varied land management and vegetation cover, *Niger. Agric. J.*, **7**, 120–123.

Kowal, J., 1970b, The hydrology of a small catchment basin at Samaru, Nigeria. IV. Assessment of soil erosion under varied land management and vegetation cover, *Niger. Agric. J.*, **7**, 134–147.

Lal, R., 1976a, Influence of residue mulches and tillage methods on soil structure and infiltration rate, In 'Modification of soil structure', ISSS Meeting of Comm. I, Adelaide, Aug. 23–27, 1976. Wiley, London.

Lal, R., 1976b, Soil erosion investigations on an alfisol in southern Nigeria, IITA Monograph No. 1, IITA, Ibadan.

Lawes, D. A., 1966, Rainfall conservation and the yields of sorghum, and groundnuts in Northern Nigeria, *Samaru Res. Bull.*, No. 70.

Macartney, J. C., P. J. Northwood, M. Dagg and R. Dawson, 1971, The effect of different cultivation techniques on soil moisture conservation and the establishment and yield of maize at Kongwa, Central, Tanzania, *Trop. Agric.* (*Trinidad*), **48**(1), 9–23.
Prentice, A. N., 1946, Tie-ridging, *East Afr. Agric. J.*, **12**, 101–108.
Wilkinson, G. E., 1975, Rainfall characteristics and soil erosion in the rainforest area of Western Nigeria, *Exp. Agric.*, **11**, 247–255.
Wilkinson, G. E., and P. O. Aina, 1976, Infiltration of water into two Nigerian soils under secondary forest and subsequent arable cropping, *Geoderma*, **15**, 51–59.

4.2

Engineering Practices to Control Erosion

W. C. MOLDENHAUER and C. A. ONSTAD

Both cultural and mechanical erosion practices in the United States are based on data collected since 1928, when the original soil erosion experiment stations were established. Data from these stations were finally collated at the Agricultural Research Service Runoff and Soil Loss Data Center at Lafayette, Indiana, in 1954. This Center developed the Universal Soil Loss Equation, which is used throughout the US and some other countries to develop criteria for controlling field erosion.

Design criteria for mechanical practices have been gradually developed by blending the research and experience of the Soil Conservation Service, the Agricultural Research Service, and the US State Agricultural Experiment Stations. These criteria, like those of the universal soil loss equation, are constantly changing with new developments or as conditions necessitate changes.

Generally, US erosion control methods have been based on use of the simplest and least costly system that is still effective for the desired cropping intensity. As land slope increases erosion control becomes more complex, especially with the new additional emphasis on environmental concerns, such as washoff of plant nutrients and pesticides (Laflen and Moldenhauer, 1971). In this chapter, we shall discuss methods of predicting soil losses and various methods and levels of erosion control, utilizing appropriate engineering designs, or management methods.

4.2.1. SOIL LOSS PREDICTION AND MODELLING

Since the late 1950's, soil loss predictions for designing conservation practices have been made using the Universal Soil Loss Equation (Wischmeier and Smith, 1965). This equation incorporates weather-based, soil, and management variables to estimate average annual field soil losses over cropping sequences. Voluminous amounts of data have been collected for refining and extending the use of this equation across the United States. At the moment, the Soil Conservation Service is using it in nearly all states east of the Rocky Mountains.

The equation is

$$A = RKLSCP$$

where the terms are as defined earlier (Chapter 3.3). All these factors have been evaluated and published for many conditions throughout the United States (Wischmeier and Smith, 1965). Recent developments have included equation modifications to take in erosion from construction sites (Wischmeier and Meyer, 1973), undisturbed areas (Wischmeier, 1974) and areas in the western United States (McCool *et al.*, 1974) not previously included.

With the advent of legislation for maintaining environmental quality, including control of erosion and sedimentation, soil erosion prediction techniques have been broadened. No longer are long-term estimates from single fields adequate. Instead, predictions are required for entire watersheds, both large and small, for long time periods, and for single-storm events. Besides the traditional parameters involved, reliable estimates of watershed erosion and sediment yield must be provided (Onstad and Moldenhauer, 1975).

Several models have evolved designed to estimate watershed erosion for single-storm events (Williams, 1975; Onstad and Foster, 1975; Snyder *et al.*, 1974). These models introduce an interrelationship between erosion and hydrologic and hydraulic parameters which greatly influence single-storm erosion events. Kling and Olson (1974) have introduced a model predicting long-term sediment yields from large watersheds. All these models are in various stages of verification and testing.

4.2.2. TILLAGE AND CONTOURING

The effectiveness of contouring decreases as slope steepness and length increase. Certain tillage practices, like listing (ridge-furrow), are very effective on the contour and may reduce soil losses by one-half. Tied ridging (Plate 7) is another effective practice (Hudson, 1971). However, there is a danger with both listing and tied ridging: if water breaks over, erosion can be much more serious than without listing, and deep rills or small gullies can occur (Plates 8 and 9). This danger can be minimized by restricting the practice to less than 3–4 per cent slopes and by adding terraces on steeper slopes.

Surface residues very effectively control erosion (Plate 10). Control is related to the quantity of residue (Table 4.6) and also its distribution over the soil surface (Wischmeier, 1975). Like other practices, if its control capability is exceeded, supplementary practices will be required.

A very promising planting system gaining popularity in the southeastern US is no-tillage sod planting in killed meadow or small grain stubble. This system has been made possible by chemicals that kill on contact the meadow top-growth and other chemicals that retard or eliminate further growth until the crop is well established. These systems not only virtually eliminate soil erosion but also increase yields. Phillips and Young (1973) presented data from Ohio, where

Table 4.6. Influence of quantity of crop residue on estimated rainfall erosion with chisel-plough and strip-tillage systems. For central Illinois, USA, Tama silt loam, 6%, 91 m, continuous corn. From Oschwald (1974)

Residue quantity (kg/ha)	Estimated rainfall erosion: Chisel-plough systems (tonnes/ha/yr)	(%)	Strip-till systems (tonnes/ha/yr)	(%)
0*	31	100	31	100
1120–2240	22	71	29	93
2240–3360	16	50	18	57
3360–4480	9	29	13	43
4480–6725	4	14	9	29
More than 6725	2	7	7	21

* Fall moldboard plough, crop residues ploughed under.

erosion from no-tillage, dead sod on the surface was 8 tonnes/ha as compared with 37 and 47 tonnes/ha for the conventional plough–disk–harrow system, cultivated and uncultivated, respectively. They showed 0 to 40 per cent increases in maize yield for no-tillage sod planting over the conventional plough–disk–harrow system (Plates 10 and 11; courtesy Dr. C. Larson).

4.2.3. TERRACES

A terrace system's main purpose is to shorten the slope length and to remove water at a velocity that will not cause erosion. There are several different kinds of terraces, and each is designed to fit a certain set of conditions.

Graded terraces are used under rainfall and soil conditions where runoff exceeds the moisture storage capacity of the soil. Grade must be sufficient to prevent ponding. Suggested minimum grades are 0·2 per cent for low permeability soils and 0·0 per cent for high permeability soils. Channel grade must not be too large or the channel will erode. The maximum grade depends on the length of run (Beasley, 1972).

Terraces may be broad-based so that crops can be planted and farm machinery operated over the entire terrace. For this type of terrace maximum land slope is about 12 per cent. As land slope increases above 5 or 6 per cent, making broad-based terraces parallel becomes increasingly more difficult because of the cutting and filling involved. The steep-backslope terrace is constructed with a 2 to 1 or steeper backslope and seeded to grass. This terrace is adapted to steeper land where farming the backslope of a broad-based terrace would be difficult. Flat channel and bench terraces are used in lower rainfall areas mainly for water conservation.

One US innovation is to farm across contour broad-based terraces rather than parallel to them. To accomplish this without excessive erosion, the terraces must be maintained and spacing must not exceed the slope length allowable for non-contour operation.

Terraces have undergone considerable evolution in the US. As the size of machinery increased and timely planting became critical, farmers could not use the contour terrace of the 1930's, with its many short rows. Also, increasing land prices made farmers reluctant to leave large areas of land in waterways, and grass-killing herbicides reduced stands and made them less effective and harder to maintain (Plate 12).

These changes have resulted in parallel terraces (as straight as possible), pushed up from the bottom side with permanently vegetated backslope and underground tile outlets with restrictive orifice plates that slowly remove water. The terraces are designed to temporarily store runoff from a 6-hr storm of 1 in 10 yr occurrence (5 cm in the midwestern United States). Outlets are designed to drain this water in 36 hr. Breathers are installed to relieve pressure if the system becomes overloaded. After the terraces have settled and the backslope grass has become well established, overtopping can be considerable without damage. The terrace's capacity must be maintained to prevent overtopping from storms of unusual occurrence. Table 4.7 shows examples of erosion from these terraces.

Table 4.7. Soil losses from grass backslope, tile outlet terraces in Iowa. From Laflen *et al.* (1972)

Location	Estimated sediment delivery to ponding area (kg/ha)	Soil loss (kg/ha)
Eldora	16,390	658
Charles City	6750	358
Creston	35,100	689
Guthrie Center	7020	872

4.2.4. GRASSED WATERWAYS AND UNDERGROUND OUTLETS

Grassed waterways can be used as terrace outlets or to stabilize an eroding drainageway from a field or farm. Their cross-sectional shape may be trapezoidal, parabolic, or V-shaped. The parabolic shape is usually used in natural depressions and the trapezoidal for terrace outlets on ridges or side slopes (Beasley, 1972).

Good conservation practices usually combine good vegetative practices with proper land use. However, under certain conditions, these practices do not adequately control water concentrations alone without scouring. In these cases, channel gradients must be reduced to maintain velocities below the erosive level, which is accomplished by conservation structures designed to dissipate the energy of the falling water. These structures vary in design but may consist of a principal spillway, an emergency spillway, and an earth embankment.

The principal spillway may be a pipe, a drop, or a chute, depending upon the amount of head to be dissipated and local considerations. Emergency spillways are designed to carry flow at a non-erosive velocity around a structure to a safe point of release. Earth embankments are used for constructing farm ponds and grade stabilization structures.

Grassed waterways, especially for terrace outlets, are becoming less popular in the US because (1) they monopolize land that could otherwise be used to produce valuable crops; (2) they require that the farmer raise planting equipment when crossing them, which slows him down and increases the chances of planter malfunction; and (3) their maintenance problem is increasing as the use of grass-killing herbicides increases. These factors have made the use of underground outlets more popular, even though they may cost more initially.

Surface inlets for underground conduits are placed in a natural depression to temporarily drain runoff stored by a terrace or embankment. With detention storage, only a small diameter conduit is needed, which can be made of clay, concrete, plastic, iron, steel, aluminium, or any other material substantial enough to withstand the weight of earth and farm machinery (Beasley, 1972).

Several other conservation practices are used in the US, like graded rows, cross-slope channels, diversions and debris basins. Descriptions and specifications for these and all mechanical practices mentioned in this chapter can be found in Beasley (1972), Schwab *et al.* (1966) and in various engineering handbooks, including one kept current by the US Department of Agriculture, Soil Conservation Service.

NOTES

1. This chapter is a contribution from the North Central Soil Conservation Research Center, North Central Region, Agricultural Research Service, USDA, Morris, Minnesota, in cooperation with the Minnesota Agricultural Experiment Station, Sci. Jour. Series No. 9090.

REFERENCES

Beasley, R. P., 1972, *Erosion and Sediment Pollution Control*, Iowa State University Press, Ames, Iowa.

Hudson, N., 1971, *Soil Conservation*, Cornell University Press, Ithaca, N.Y.

Kling, G. F., and G. W. Olson, 1974, The sediment transport computer model, Cornell Agronomy Mimeo 74–11, Cornell University, Ithaca.

Laflen, J. M., H. P. Johnson and R. C. Reeve, 1972, Soil loss from tile-outlet terraces, *Jour. Soil and Water Conserv.*, **27**, 74–77.

Laflen, J. M., and W. C. Moldenhauer, 1971, Soil conservation on agricultural land, *Jour.* Soil and Water Conserv., **26**, 225–229.

McCool, D. K., W. H. Wischmeier and L. C. Johnson, 1974, Adaptation of the universal soil loss equation to the Pacific Northwest, Paper No. 74–2523, presented at 1974 Winter Meeting of ASAE, Dec. 10–13, Chicago, Ill.

Onstad, C. A., and G. R. Foster, 1975, Erosion modeling on a watershed. *Transactions of the American Society of Agricultural Engineers*, **18**(2), 288–292.

Onstad, C. A., and W. C. Moldenhauer, 1975, Watershed soil detachment and transportation factors, *Jour. of Environmental Quality*, **4**(1), 29–32.

Oschwald, W. R., 1974, In *Conservation Tillage, a handbook for farmers*, Soil Conservation Society of America, Ankeny, Iowa.

Phillips, S. H., and H. M. Young, Jr., 1973, *No-Tillage Farming*, Reiman, Milwaukee, Wisconsin.

Schwab, G. O., R. K. Frevert, T. W. Edminister and K. K. Barnes, 1966, *Soil and Water Conservation Engineering*, Wiley, New York.

Snyder, W. M., R. R. Bruce, L. A. Harper, A. W. Thomas and R. A. Leonard, 1974, A model for runoff of pesticides from small upland watersheds. Presented at Soil Science Society of America meeting, Nov. 10–15, Chicago, Ill.

Williams, J. R., 1975, Sediment yield prediction with universal equation using runoff energy factor, *Proceedings, 1972 Sediment Prediction Workshop*, US Government Printing Office, Washington, D.C., in Press.

Wischmeier, W. H., 1974, New developments in estimating water erosion, *Proceedings of the 29th Annual Meeting Soil Conservation Society of America, August 11–14, 1974, Syracuse, New York*, SCSA, Ankeny, Iowa.

Wischmeier, W. H., 1975, Estimating the soil-loss equation's cover and management factor for undisturbed areas, *Proceedings, 1972 Sediment Prediction Workshop*, US Government Printing Office, Washington, D.C., in Press.

Wischmeier, W. H., and L. D. Meyer, 1973, *Soil erodibility on construction areas*, Highway Research Board, National Academy of Science Special Report 135, Washington, D.C., pp. 20–29.

Wischmeier, W. H., and D. D. Smith, 1965, *Predicting rainfall-erosion losses from cropland east of the Rocky Mountains—Guide for selection of practices for soil and water conservation*, Agricultural Research Service, US Department of Agriculture, Agriculture Handbook No. 282.

4.3

Soil Management Systems and Erosion Control

R. LAL

It is clear from the foregoing chapters that in spite of the major importance of soil erosion in the humid tropics, very little information is available to use in designing appropriate cropping systems for such areas. The Universal Soil Loss Equation provides a framework potentially useful for designing appropriate systems, once values for the factors in the equation have been derived. In this chapter the crop management factor, *C*, will be discussed in the light of information recently obtained on soils at the International Institute of Tropical Agriculture at Ibadan, in southern Nigeria. Their management will be discussed in relation to slope length and gradient (the *S* term).

4.3.1. SLOPE GRADIENT *S*

Slope gradient is important because the velocity of water movement and its ability to detach and carry soil particles increases exponentially with increase in slope angle.

If this is expressed as

$$\text{Soil loss} = C's^{K} \tag{4.1}$$

where C' and K are constants while S is the slope in per cent, under the bare fallow the exponent K ranges from 1·1 to 1·2. The water runoff, however, is more dependent on soil hydrologic characteristics and less on the slope steepness (Lal, 1976).

Water runoff and soil loss observations under straw mulches have supported the belief that slope steepness is not the dominant factor when mulch-farming techniques are employed to prevent erosion. The data in Table 4.8 indicate that soil loss was not significantly increased when slope steepness increased from 1 to 15 per cent with mulch rates of 4 to 6 tonnes of straw/ha. Fortunately, therefore, soil management can be more important than the slope gradient as a factor determining potential erosion hazard. It is difficult and sometimes impossible to alter the slope, but proper management practices can reduce erosion to negligible levels.

Table 4.8. Effect of slope steepness on soil loss (tonnes/ha) under continuous maize with mulch compared with the loss from bare soil IITA, Ibadan, Nigeria. Rainfall = 105·4 mm

Slope (%)	Bare plot	Maize–Maize (mulched)
1	0·8	0·00
5	4·27	0·01
10	4·27	0·09
15	29·80	0·02

4.3.2. CROPPING MANAGEMENT

There is enough evidence in the literature to indicate that good farming methods are more important to minimizing erosion than the nature or canopy cover of the crops to be grown. For example, soil loss from close-canopy cowpeas can be as high as under maize in the early growth stages. On the other hand, soil loss under maize, an open-row crop, can be significantly lower than that under cowpea, if the former is grown with mulch-farming techniques. Indigenous farming methods utilizing mixed and relay cropping with small plot sizes and minimal cultivation which leaves the soil little exposed to direct erosive action, are common features in the humid tropics. Obviously vigorous well grown crops are also necessary to ensure a good soil cover. In this chapter the principles of soil management necessary to minimize runoff and soil loss under arable farming, will be discussed.

4.3.2.1. Mulch and Mulch-Farming Techniques

As discussed in the previous chapter, soil management methods employed in preventing water runoff and soil loss include strip-cropping, tied ridging and minimum or no-tillage techniques. These are all designed to maintain the infiltrability of the soil and decrease runoff and soil loss. A mulch prevents surface sealing by preventing direct raindrop impact on the soil, and by encouraging enhanced biological activity, which leads to the development of macropores in the soil. An example of runoff and soil loss under different mulch rates is given in Table 4.9.

The soil and water loss data shown in Table 4.9 are an average of four slopes of 1, 5, 10 and 15 per cent. The water runoff and soil loss decreased exponentially with increase in mulch rate:

$$\text{Soil loss} = C''M^{-K} \qquad (4.2)$$

where M is the mulch rate and C'' and K are constants.

Table 4.9. Effect of mulch rate on runoff and soil loss on uncropped land at IITA, Ibadan, Nigeria. Rainfall = 61·1 mm

Mulch rate (tonnes/ha)	Runoff (%)	Soil loss (tonnes/ha)
0	50·0	4·83
2	19·7	2·48
4	8·0	0·52
6	1·2	0·05

Under Ibadan conditions, on the Egbeda soil, *K* ranges between 0·06 and 0·30. It is evident that within the slope range of 1–15 per cent, runoff and soil loss are negligible with mulch rates of 4–6 tonnes/ha. The exponential slope–soil relationship does not hold true for the mulched plots.

4.3.2.2. No-Tillage System and Soil Conservation

One of the means of procuring mulch on the soil surface is from crop residues left on the surface when using minimum or no-tillage techniques. Weed control, one of the major objectives of primary tillage, can be obtained by the use of herbicides. Using crop residues as a surface mulch also reduces weed growth. A no-tillage system has the further advantage of moisture conservation in the soil profile, in addition to decreasing runoff and reducing soil loss to a minimum (Table 4.10) (Plates 13, 14, 15 and 16).

Table 4.10. No-tillage effects on runoff and soil loss under cowpeas at two weeks after planting at IITA, Ibadan, Nigeria. Rainfall = 63·0 mm

	Runoff (%)		Soil loss (tonnes/ha)	
Slope (%)	Bare ploughed	No-till	Bare ploughed	No-till
1	69·8	1·3	1·0	0·00
5	71·9	1·6	2·6	0·00
10	74·0	3·2	19·6	0·01
15	87·6	2·7	33·3	0·00

Once again the exponential slope–soil loss relationship does not exist for the no-tillage system.

In addition to the principal advantages of no-tillage systems in soil and water conservation, there are some other advantages. For instance, high soil temperature can be one of the major limiting factors in arable crop production in the tropics. Soil temperature differences of as much as 10 °C in the surface 5-cm depth have been observed in no-till plots with crop residue mulches, as compared with ploughed plots, where temperatures rise to as high as 45 °C. The organic matter content of the surface horizon in no-till systems is better maintained and so is the water-holding capacity of the soil. The moisture retention characteristics of the soil are also changed in favour of no-tillage treatments. The surface soil has higher infiltration rates and the saturated hydraulic conductivity is also high (Table 4.11).

Table 4.11. No-tillage effects on the saturated hydraulic conductivity (cm/sec) for different soil layers. IITA, Ibadan, Nigeria Egbeda series. Each figure is an average of nine replicates

Treatment	0–10 cm	10–20 cm	20–30 cm
Ploughed	0·048	0·021	0·016
No-till	0·055	0·044	0·043

There are some controversial issues to be resolved in connection with minimum tillage techniques. The controversy exists primarily because of the lack of uniformity in techniques employed for the no-tillage systems. The amount of crop residue and initial soil conditions are important parameters to be considered in the interpretation of these results. No-tillage techniques without crop residue mulches are likely to be a failure. If the ground is initially compacted, the structure has to be improved, not through deep ploughing, but by using cover crops such as *Pueraria*, *Desmodium*, *Brachiaria*, etc. A proper fertilizer regime for use in no-till crop production has also to be established and used.

The no-tillage system has a tremendous potential for the humid tropics and probably for the semi-arid regions as well. Although there are agronomic problems still to be resolved, the erosion control provided without recourse to expensive engineering methods is an overwhelming advantage. Research on the problems of weed control, fertilizer rates and placement, and the use of appropriate farm machinery, is still needed, but none of the problems appear difficult, except in drier areas where preservation of mulch material through a long dry season may not be possible.

SUMMARY

The importance of slope gradient (*S*) and crop management are discussed. The slope steepness factor can be considerably modified by the soil and crop management practices. The important factor is not which crop is to be grown, but the cultural practices adopted. Surface mulching decreases the water and soil loss to a minimum regardless of the slope within a slope steepness range of 1 to 15 per cent.

A no-tillage system of soil management, with adequate crop residue on the soil surface, ensures soil and water conservation with negligible water runoff and virtually no soil loss. The infiltration rate under the no-tillage system is maintained and saturated hydraulic conductivity is high. In addition, the higher level of soil organic matter in the no-tillage system improves the water and nutrient-retention capacity of the surface soil. The no-tillage, mulch-farming techniques have a tremendous potential for use in developing continuous arable production systems in the humid tropics.

REFERENCE

Lal, R., 1976. *Soil erosion investigations on Alfisols in southern Nigeria.* IITA Monograph No. 1.

4.4

Possibilities of Using Soil Conditioners for Soil Erosion Control

D. GABRIELS, L. MAENE, J. LENVAIN and M. DE BOODT

The Universal Soil Loss Equation

$$A = RKLSCP$$

can be reduced for laboratory tests to a simple relationship by standardizing slope length and gradient of test boxes, making $S = 1$ and $L = 1$, and assuming $P = 1$ and $C = 1$. Then:

$$A = RK$$

Results of laboratory trials at different rainfall intensities indicated a linear relationship between side-splash erosion (*SSE* in g/h) and the kinetic energy (*KE* in J/h) of the rainfall, in agreement with the results of Ekern (1951). Similar relationships exist between soil loss (*A* in g/h), amount of runoff (*W* in cm^3/hr) and the kinetic energy of the rainfall.

$$SSE = K_1 \times KE$$
$$A = K_2 \times KE$$
$$W = K_3 \times KE$$

where K_1, K_2, K_3 are, respectively, the detachability (stability of the aggregated soil surface), the erodibility and the infiltrability. These linear relationships provide convenient criteria for assessing to what extent these properties can be improved by use of soil amendments.

If unamended and amended soils are compared, the effect of the amendment on the *K* value and subsequently the splash erosion, the soil loss and the runoff can be obtained:

$$SSE = K_1'K_1 \times KE$$
$$A = K_2'K_2 \times KE$$
$$W = K_3'K_3 \times KE$$

where K_1', K_2', K_3' are, respectively, splash erosion reduction factor, soil loss reduction factor (also called erosion reduction factor) and runoff reduction factor. For an untreated soil, $K' = 1$. To the extent that soil conditioning reduces erosion, K' will be <1. A general index table can be constructed.

K′-value	*Amendment Evaluation*
1	Poor
0·8–1	Fairly poor
0·6–0·8	Fair
0·4–0·6	Good
0·2–0·4	Very good
0–0·2	Excellent

4.4.1. SOIL CONDITIONING

In order to prevent the breakdown of soil aggregates from raindrop impact, their aggregate stability should be improved. Soil conditioners can provide the soil surface and even the upper soil layer with a more stable structure and with optimal physical properties, minimizing aggregate destruction and helping to preserve a high infiltration capacity.

A soil conditioner can be applied in two different ways: (1) by surface treatment; (2) by incorporation. In surface treatments the conditioner is sprayed, at its optimal dilution rate, as a film (mulch) over the dry soil surface. For incorporation, the conditioner, at its proper dilution rate, is mixed with the soil at different initial soil moisture contents. Studies by De Boodt (1970) and De Boodt and De Bisschop (1970) showed the existence of an optimal initial or total moisture content for conditioning (Rigole and De Bisschop, 1972; Vandevelde and De Boodt, 1972; Gabriels and De Boodt, 1972). Examples of the effects of various conditioners on erosion characteristics are given in this chapter.

4.4.2. MATERIALS AND LABORATORY METHODS

The particle size distributions of the soils used, before treatment with the soil conditioners, are given in Table 4.12.

Table 4.12. Mechanical analyses of the soils studied

	% > 50 μm	% (2–50 μm)	% (0–2 μm)
Sand (dune sand)	97·9	1·6	0·5
Heavy sandy loam	18·6	64·0	17·4
Heavy loam	10·2	65·9	23·9
Sandy clay	55·1	21·1	23·8

The dune sand and the clayey aggregates were placed to a uniform depth of 5 cm in a soil pan 30 cm × 30 cm. Free drainage was provided by means of a fritted false bottom. Simulated rainfall was applied to the pans adjusted to 33 per cent slope using the waterdrop applicator described by Gabriels *et al.* (1973). The rain intensities ranged from 50 to 65 mm/h during a 1-hr run, corresponding to a kinetic energy of about 1100–1200 J/m^2 J/m^2 hr. Determinations were made of the amount of side-splash erosion, the runoff water and total soil loss after one hour of rainfall.

The soil conditioners used are listed in Table 4.13. (Trade names and company names are included for the benefit of the reader and do not form an endorsement of preferential treatment of the product by the University of Ghent.)

Table 4.13. Chemicals used for soil stabilization in the laboratory

Basic compound	Form	Trade name	Company
Polyvinyl acetate	Emulsion	Curasol AE Curasol AH	Farbwerke Hoechst, Germany
Polyacrylamide	Solution	Humofina PAM	Labofina, Belgium
Polyvinyl pyrrolidone	Solution		
Asphalt	Emulsion	Humofina bitumen	Labofina, Belgium
Polyvinyl alcohol	Solution		
Polyurethane	Solution	PRB 2006	N.V. PRB, Belgium
Polyethyleneglycol	Solution		
Latex	Emulsion	Petroset SB	Phillips Petroleum Company (USA)
Latex + bark			

4.4.3. RESULTS AND DISCUSSION

4.4.3.1. Surface Treatment

Table 4.14 contains the results (K_1'-value) of the tests on dune sand treated with mulches of soil conditioners.

Table 4.14. Evaluation of soil-conditioner mulches on dune sand

Basic compound	Active material (g/m^2)	Diluted amount (l/m^2)	K_1' (*SSE*)	Evaluation
Polyvinyl acetate	20–30	0·65–0·67	0–0·4	Very good–excellent
Polyacrylamide (+ glyoxal)	10–20	1	0–0·4	Very good–excellent
Asphalt	17·5	5·25	0	Excellent
Latex	32·5–55	1·8–2·25	0–0·2	Very good–excellent
Polyurethane	25	0·75	0–0·2	Excellent
Latex + bark	27·5 + 90	2·25	0–0·2	Very good–excellent

A treatment with asphalt or latex emulsion is effective at high dilution rates so that the soil conditioner can penetrate the sand instead of forming a hard concentrated film on the surface. When bark is mixed with latex, the amount of latex can be reduced to the amount sufficient to retain the bark on the slope.

Table 4.15 contains the results (K_1', K_2', K_3'-values) of the tests on clayey soils. In addition to the tests on the dune sand, where only splash erosion (*SSE*) could be measured, for clayey soils the amounts of runoff water (*W*) and wash erosion (*A*) were also measured. The erosion reduction factor K_3', seems to be the most interesting and the most critical with respect to soil loss (*E*).

4.4.3.2. Incorporation Treatment

When hydrophobic soil conditioners such as asphalt or latex emulsions are applied as surface treatments in concentrated form (not highly diluted), an impermeable film is usually formed on the soil surface. When hydrophobic soil conditioners are mixed with the soil, erosion is reduced only when sufficiently large aggregates and pores are formed so that water can infiltrate. For dune sand, the total moisture content (initial moisture + moisture added with the soil conditioner) was kept at 20 per cent, and for clayey soil at 25 per cent. Table 4.16 contains the K_1' values of the tests on dune sand mixed with soil conditioners.

Table 4.16. Evaluation of incorporation treatments on dune sand

Basic compound	Active material (% on dry soil weight)	K'_1 (*SSE*)	Evaluation	Runoff (% of total rainfall)
Polyvinyl acetate		0–0·6	Good–excellent	
Polyacrylamide	2 (M.W. 150,000)	0·8–1	Poor	
Polyacrylamide + cross linker	2 (M.W. 150,000)	0–0·2	Excellent	
Asphalt (1)*	7·5	0·8–1	Poor	89
Asphalt (2)†	7·5	0·2–0·4	Very good	13
Asphalt (3)‡	7·5	0·2–0·4	Very good	0

* The formed aggregates were air dried, passed through a 5·35 mm sieve and placed in the metal pan of the rainfall simulator.

† The air-dry aggregates as such, without sieving, were poured in the metal pan.

‡ The treated wet aggregates were placed in the metal pan and allowed to dry *in situ*.

When asphalt was mixed with the sand, runoff was sometimes increased The hydrophobic emulsion made the aggregates water-repellent by increasing the liquid–solid contact angle. This plays a role only when the aggregates are sufficiently small (smaller than 5·35 mm, as with treatment (1), where 89 per cent of the total rainfall was lost as runoff). If the aggregates are not broken down, larger pores occur between the aggregates, allowing the water to infiltrate in spite

Table 4.15. Evaluation of soil conditioners applied as surface treatments on clayey soils

Basic compound	Active material (g/m^2)	Diluted amount (l/m^2)	K'_1 (SSE)	Evaluation	K'_2 (W)	Evaluation	K'_3 (E)	Evaluation
Polyvinyl acetate	25	0·65	0·6–0·8	Fair	0·6–1	Fair	0·2–0·4	Very good
Polyacrylamide	8–20	1	0·2–0·6	Good–very good	0·2–0·6	Good–very good	0–0·2	Excellent
Polyurethane	10–15	1	0·6–0·9	Fair–fairly good	0·6–0·9	Fair–fairly poor	0·2–0·4	Very good–excellent
Polyurethane	20	1·5	0–0·3	Excellent–Very good	0·8–1	Fairly poor	0	Excellent
Latex	27·5	2·25	0·7	Fair	1	Poor	0·85	Fairly poor
Latex	55	2·25	0·55	Good	0·65	Fair	0·15	Excellent
Latex + bark 2 cm	27·5 180	2·25	0·50	Good	0·8–0·9	Fairly poor	0·40–40·5	Good
Latex + bark 8 mm	27·5 180	2·25	0·35	Very good	0·6	Fair–good	·0·15	Excellent

Table 4.17. Evaluation of incorporation treatments on clayey soils

Basic compound (MW = approx. molecular weight)	Active material (%) (on dry soil weight)	K'_1 (SSE)	Evaluation	K'_2 (W)	Evaluation	K'_3 (E)	Evaluation
Polyacrylamide + glyoxal MW 150,000	1	0·6–0·8	Fair–	0·4–0·6	Excellent	0–0·2	Excellent
	2	0–0·2	excellent	0			Excellent
Polyacrylamide MW c. 10^6 (anionic)	0·2	0·6–0·8	Fair	0·8–1	Fairly poor	0·8–1	Good–
	0·4	0·4–0·6	Good	0·8–1	Fairly poor	0·2–0·6	very good
Polyvinyl alcohol MW 100,000	1	0·8–1	Fairly poor	0·4–0·6	Good	0–0·2	Excellent
	2	0·2–0·4	Very good	0–0·2	Excellent	0–0·2	Excellent
Polyvinyl pyrrolidone MW 100,000	0·5	1	Poor	0·4–0·6	Good	0·2–0·4	Very good
	1	0·8–1	Fairly poor	0·2–0·4	Very good	0·2–0·4	Very good
Polyvinyl acetate	2·5	0·4–0·8	Fair–good	0·6–1	Fair–fairly poor	0·2–0·4	Very good
Asphalt	7·5	0·4–0·6	Good	0·2–0·6	Good–very good	0·2–0·4	Very good
Asphalt (1)*	7·5	0·4–0·6	Good	1	Poor	0·8–1	Fairly poor

* The formed aggregates were aid dried, passed through a 5·35 mm sieve and placed in the metal pan of the rainfall simulator.

of the high liquid–solid contact angle. Polyacrylamide solutions produced a high infiltrability for water whether large or small aggregates were formed.

Table 4.17 contains the K_1', K_2', and K_3' values of the tests on clayey soils mixed with soil conditioners. Here also, when asphalt was mixed with the soil and only small aggregates were formed (5·35 mm), runoff and subsequent soil losses were high. A bitumen emulsion, because of its hydrophobic properties, is unsuitable for surface applications, as well as incorporation treatments. The erosion can be reduced only when large aggregates and pores are formed so that water can infiltrate.

4.4.4. LABORATORY EXPERIMENTS ON TROPICAL SOILS IN WESTERN MALAYA

Rainfall simulator experiments were set up to study erosion phenomena on some tropical soils in Western Malaya. The soils are classified locally as Serdang, Rengam and Durian Series (Table 4.18).

Table 4.18. Particle size distribution of some Western Malaya soils

Soil	% clay	% silt	% sand
Serdang	19·2	3·8	77
Rengam	30·6	10·2	59·2
Durian	47·8	14·0	38·2

Rainfall data for the period from 1929 until 1972 show that the maximum intensities for a rain duration of 30 minutes varied between 82·0 mm/hr and 160 mm/hr, and for a duration of one hour between 58 mm/hr and 114 mm/hr.

The following treatments were applied to the soils:

T_1: untreated air-dry soil aggregates passed through an 8 mm sieve,

T_2: soil mixed with asphalt emulsion (7·5 per cent active material), allowed to dry, and passed through the 8 mm sieve,

T_3: soil mixed with asphalt emulsion (7·5 per cent active material), allowed to dry, not sieved.

The air-dry soil aggregates were placed in the metal pans of the rainfall simulator, tilted to a slope of 5 per cent and subjected to an artificial rainfall of about 115 mm/hr for 30 minutes. The runoff and soil loss (Figure 4.2) increased drastically for the fine asphalt-treated soil fraction <8 mm. The hydrophobicity apparently prevented water entry. When the whole soil was used, water penetrated the large pores between the hydrophobic aggregates, and runoff and erosion were reduced.

4.4.5. FIELD EXPERIMENTS ON TROPICAL SOILS IN JAVA

Erosion tests were conducted in the field on two soil types in Java, at Lembang on an 'Andosol' and at Jonggol on a 'Red-yellow podsolic soil'. The following

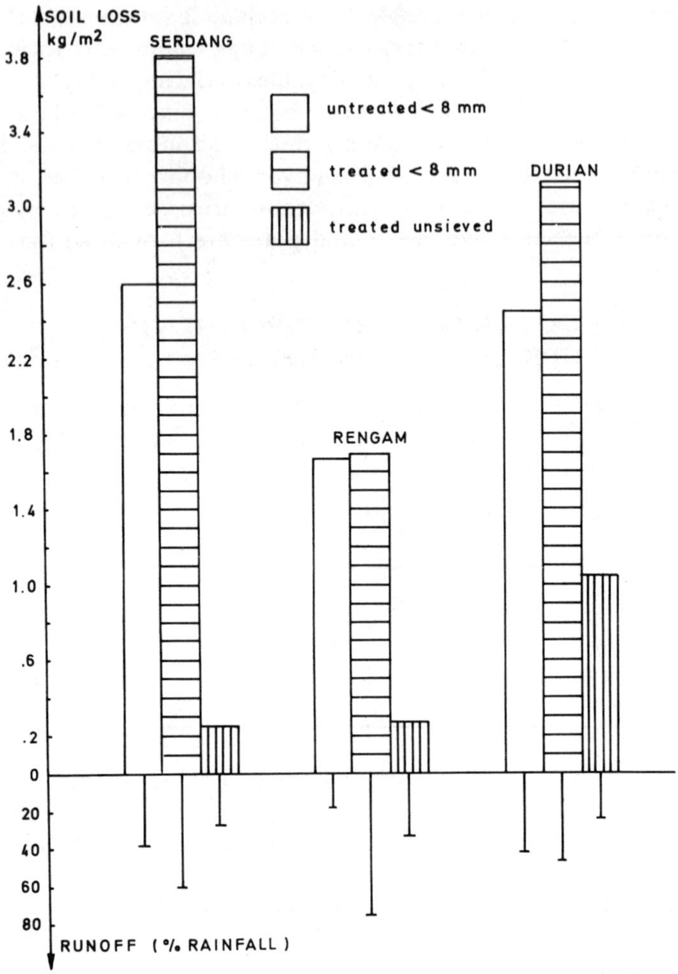

Figure 4.2. Soil loss and runoff from Malaysian soils treated with an asphalt emulsion

treatments were used: (1) control, (2) an asphalt emulsion at 1 l/m^2 incorporated to about 10 cm, (3) an 8 per cent polyacrylamide solution at 1 l/m^2 incorporated to about 10 cm, (4) organic mulch of dry grasses at 0·5 tonne/ha.

The field plots were 6 m × 2 m, on a 13 per cent slope. Eroded soil and runoff were collected in the first months of 1972. For the untreated plot, a linear relationship ($y = ax + b$) was found between the cumulative soil loss (y) and the total rainfall or the erosivity index $EI_{30}(x)$, for the treated plots the regression between the cumulative soil loss (y) and the cumulative EI_{30} is of the form $y = a_1x^2 + a_2x + a_3$. The EI_{30}-value, the factor most closely related to erosion, was calculated after Wischmeier *et al.* (1958).

When the gradient of the curve for the treated plot becomes identical to that of the untreated plot, then the effectiveness of the treatment disappears. The corresponding EI_{30} value can be calculated as follows:

$$x = \frac{a - a_1}{2a_2}.$$

For the bitumen-treated plot, the maximum $EI_{30} = 447$, corresponding to a total rainfall of 635 mm; for the polyacrylamide-treated plot it is $EI_{30} = 1528$, or a total rainfall of 2639 mm; for the organic mulch it is $EI_{30} = 2510$, or a total rainfall of 4319 mm.

The bitumen treatment is only effective for less than one rain-period and therefore, because of its water-repellency, not always suitable for soil conservation purposes. The polyacrylamide treatment is more effective, but also more expensive. For agricultural purposes, a soil conditioner treatment in strips where, for example, sugar beet seeds are planted can solve the economic problems. The higher effectiveness of the organic mulch is due to the fact that it acts as an energy-breaker of the falling raindrops.

SUMMARY

Synthetic soil conditioners were evaluated in the laboratory using a rainfall simulator. The criteria used were the rates of side-splash erosion, runoff water and soil loss per unit kinetic energy of rainfall for the treated soils compared to the untreated. The hydrophobic soil conditioners such as the asphalt emulsions were only effective in decreasing the amounts of runoff water and soil loss when incorporated into the upper soil layer provided that sufficiently large aggregates were formed, with large pores in between, allowing water to infiltrate. Field experiments on tropical soils in Java also proved the low effectiveness of the asphalt emulsion treatment compared to the polyacrylamide treatment and the organic mulch. For greater effectiveness and for economic reasons attention should concentrate in the future on the use of mulches, or possibly of mixtures of organic matter (peat, bark, straw, compost) with soil conditioners.

REFERENCES

De Boodt, M., 1970, New possibilities for soil conditioning by means of diluted bituminous emulsions, FAO, ECA, W.R./70/4/(b)11, 15 May 1970.

De Boodt, M., and F. De Bisschop, 1970, Bodemstabilisatie en strukturatie met behulp van geemulgeerde bitumen en polymeren, *Cultuurtechnisch tijdschrift*, **9**, No. 5, 1–10.

Ekern, P. C., 1951, Raindrop impact as the force initiating soil erosion, *Soil Sci. Soc. Amer. Proc.*, **15**, 7–10.

Gabriels, D., and M. De Boodt, 1972, Theoretical and practical approach for determining the optimal moisture content at the moment of soil conditioning, *Mitteilgn. Dtsch. Bodenkundl. Gesellsch.*, **15**, 185–203.

Gabriels, D., M. De Boodt and W. Minjauw, 1973, Description of a rainfall simulator for soil erosion studies, *Mededelingen Fakulteit Landbouwetenschappen, State Univ. Ghent, Belgium*, **38** (2), 294–303.

Kirkham, D., and J. E. Adams, 1960, Apparatus for measuring erodibility, runoff, infiltration and other physical properties of soils in place, US Patent Office 2, pp. 923–1048.

Rigole, W., and F. De Bisschop, 1972, The formation of adhesive links by means of emulsions, *Mededelingen Fakulteit Landbouwetenschappen, State Univ. Ghent, Belguim, 1972*, **37** (3), 938–954.

Vandevelde, R., and M. De Boodt, 1972, Reaction of soil textures on soil conditioning with an emulsion of bitumen and a solution of adhesive (PAM), *Mededelingen Fakulteit Landbouwetenschappen, State Univ. Ghent, Belgium*, **37** (3), 993–1013.

Wischmeier, W. H., and D. D. Smith, 1965, Rainfall erosion losses from cropland east of the Rocky Mountains, USDA Handbook 282.

Wischmeier, W. H., D. D. Smith and R. E. Uhland, 1958, Evaluation of factors in the soil-loss equation, *Agricultural Engineering*, **39**, 8, 458–462.

4.5

Aggregate Size and Erosion Control on Sloping Land Treated with Hydrophobic Bitumen Emulsion

I. PLA

4.5.1. INTRODUCTION

Crust formation and erosion are common problems in the cultivated soils of the humid tropics, due to the high rainfall intensities and poor physical properties of many of the soils (Pla *et al.*, 1974). In many areas of the humid tropics with sloping or rolling surfaces it is as important to minimize crust formation as to prevent erosion, though the former influences the magnitude of the latter. To achieve this, it is necessary to keep the infiltration rate high enough to minimize runoff. Hydrophobic emulsions have been proved to be effective in stabilizing aggregates and preventing crust formation, in surface soils submitted to high rainfall intensities (De Boodt, 1972; Pla, 1975). These relatively cheap conditioners have therefore been studied as erosion preventatives. However, in the case of a sloping surface, it was found that when aggregates from a silt loam soil, passing a 2·86 mm mesh sieve, were treated with anionic bitumen emulsion and subjected to an intense storm, the runoff and soil loss could be as high or higher than in a non-treated soil. The reason appeared to be the difficulty that water has in penetrating the surface layer of hydrophobic soil aggregates. Nevertheless if the stabilized hydrophobic aggregates are large enough, water may flow through the pores between them at rates preventing runoff, without wetting the aggregates themselves. The critical size of the aggregates would presumably depend on rainfall intensity and slope gradient. The experiments reported here were aimed to test this assumption and to find if there are conditions under which it would be possible to use hydrophobic bitumen emulsions both for crust prevention and erosion control on sloping land under different rainfall intensities.

4.5.2. MATERIALS AND METHODS

A surface silt–loam soil (17·4 per cent clay 0–2 μ; 64·0 per cent silt 2–50 μ; 18·6 per cent sand > 50 μ; 1·34 per cent organic carbon; 1·93 per cent $CaCO_3$) was treated with 1·5 per cent bitumen emulsion (Humofina B 2864) following the procedure described by Gabriels and De Boodt (1972). From the air-dry treated soil, aggregates of <2 mm, 2–4·2 mm, 4·2–5·3 mm and 5·3–8·0 mm were separated by sieving. The same aggregate fractions were obtained by sieving an air-dry non-treated sample of the same soil.

Layers, 5 cm thick, of the different air-dry soil aggregates were placed in soil pans 30 cm × 19 cm, provided with a removable fritted 'false' bottom, the soil surface being levelled to correspond with the lip of a runoff concentrator attached to the front wall of the pan. Four pans in the same body allowed simultaneous tests on the four aggregate sizes. Splashboards were placed on the sides of the pans. The set of four pans could be tilted to the desired slope gradients of 0, 6, 18, and 33 per cent.

Simulated rainfall with different sustained intensities (15, 30, and 50 mm/hr) and drop size ranging from 4·5 to 5·2 mm diameter was applied for 1 hr and in some cases for 2 hrs, using the rainfall simulator described by Gabriels *et al.* (1973). Four raingauges were used to check the rainfall intensity and distribution in the position of each soil pan. Runoff water and half of the side-splashed soil material from each pan were collected every five or ten minutes. After drying, the soil material was weighed. In some tests, water drained below the layer of soil aggregates was collected and measured at regular intervals. At the end of each test the 'false' bottoms were removed together with the soil on top of them, and the depth and degree of wetting of the aggregates was carefully observed and measured.

4.5.3. RESULTS AND DISCUSSION

The cumulative values of runoff water and soil loss as a function of rainfall duration for different rainfall intensities, slopes and aggregate sizes of treated soil are shown in Figure 4.3. Results are presented in Figure 4.4 for two rainfall intensities and 33 per cent slope, using aggregates of non-treated soil. No runoff was produced after 1 hr of a 15 mm/hr rainfall in aggregates of non-treated soil. The vertical scales for the values of runoff and soil losses are related to rainfall intensities in order to get lines with lengths and slopes corresponding to the same cumulative rainfall at each time. This facilitates the independent analysis of the influences of rainfall intensity, slope gradient and aggregate size.

Analysis of Figures 4.3 and 4.4 shows that:

(a) Runoff losses from treated soil are high and reach equilibrium within a short time for all aggregate sizes and slope ranges for storms of 50 mm/hr intensity. The steady-state runoff is achieved only for aggregates of <2 and 2–4·2 mm size for 30 mm/hr intensity. Similarly, if the intensity drops to 15 mm/hr, this equilibrium runoff is attained only for aggregates of <2 mm, regardless of slope gradient. No runoff was obtained for 5·3–8 mm aggregates with rainfall of 30 and 15 mm/hr and slope gradients of 18 and 33 per cent. The same is observed for 4·2–5·3 mm aggregates with 15 mm/hr rainfall and 5 per cent slope.

Soil losses, for a given cumulative rainfall, in treated soil, are only high for

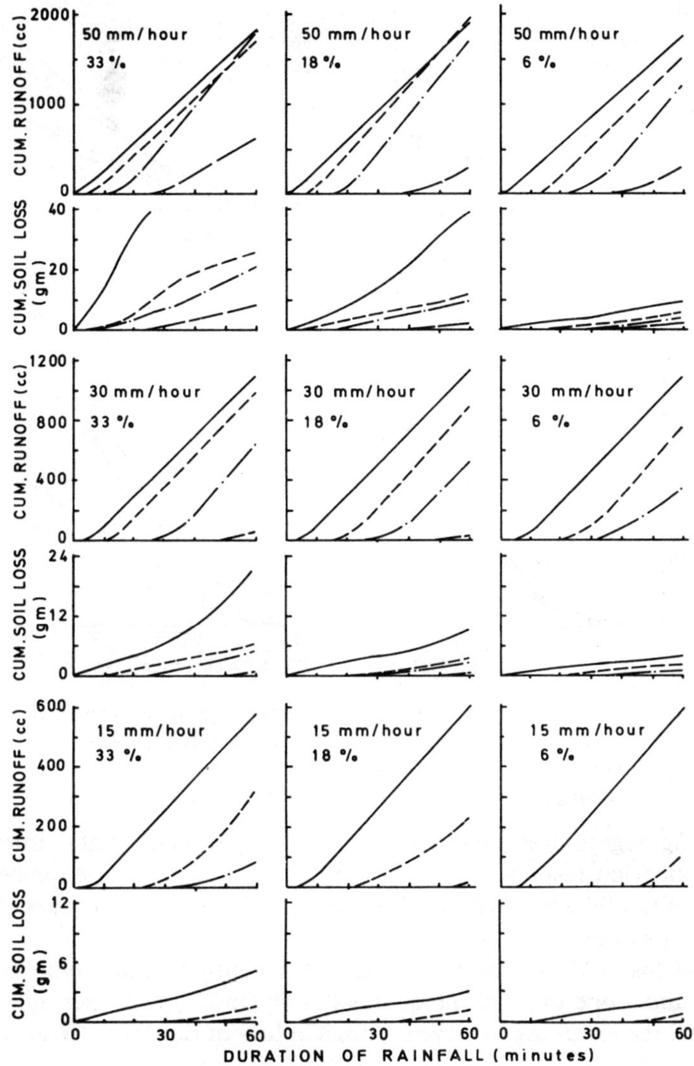

Figure 4.3. Cumulative soil loss and runoff for aggregates from a silt loam treated with 1·5 per cent of bitumen emulsion at 3 rainfall intensities and 3 slopes. Aggregate sizes: ——— <2 mm; – – – 2–4·2 mm; –·–·– 4·2–5·3 mm; ——— 5·3–8 mm

aggregates <2 mm; 2–4·2 mm and 4·2–5·3 mm with rainfall intensity of 50 mm/hr and 33 per cent slope gradient; for aggregates <2 mm with a rainfall intensity of 50 mm/hr and 18 per cent slope; and for <2 mm aggregates with a rainfall of 30 mm/hr and 33 per cent slope.

It is evident that the fraction of rainfall water lost as runoff is independent of rainfall intensity and slope for hydrophobic aggregates smaller than 2 mm. For

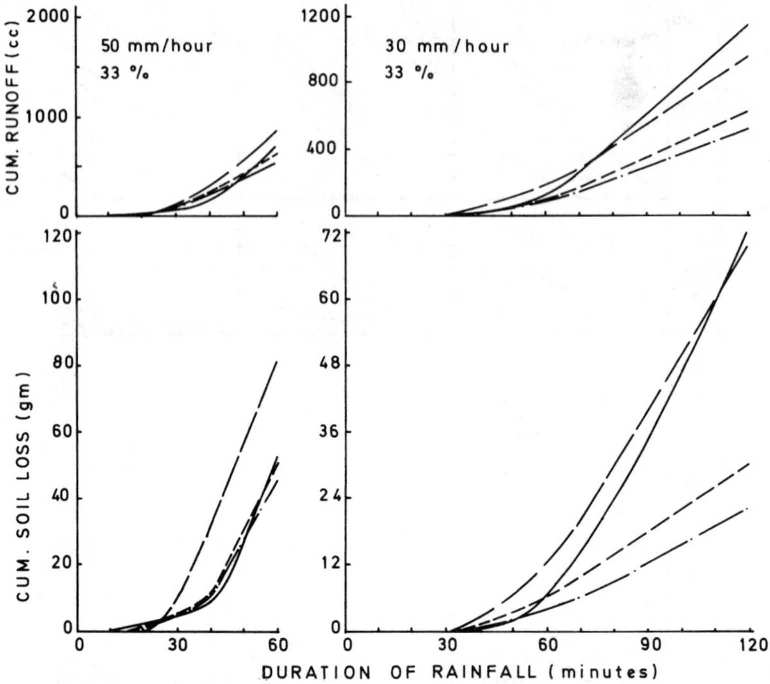

Figure 4.4. Cumulative soil loss and runoff for aggregates from a silt loam (untreated, and initially air-dry) at rainfall intensities of 30 and 50 mm/hr at a slope gradient of 33 per cent. Aggregate sizes:
—— <2 mm; --- 2–4·2 mm; -·-·- 4·2–5·3; —— —— 5·3–8 mm

the other aggregate sizes there is a critical combination of rainfall intensity and slope gradient corresponding to a sharp decrease in runoff. As expected, the influence of rainfall intensity on volume of runoff is much more marked than the influence of slope.

The soil losses from treated soil are significantly affected both by rainfall intensity and slope even in the aggregates <2 mm. For larger aggregates of treated soil the losses are low even at high values of rainfall intensity and slope gradient.

(b) Compared with the treated soil, runoff starts later in aggregates of non-treated soil for rainfall intensities of 30 and 50 mm/hr and 33 per cent slope gradient, except for 5·3–8 mm aggregates. No marked influence of aggregate size on runoff is observed. When comparison of soil losses is made for a given amount of runoff, they are seen to be much larger in the non-treated soil, the greatest difference being observed for aggregates 5·3–8 mm. However runoff is greater from the treated soil, so that erosion for a given storm is not necessarily less.

Increase in runoff and soil loss from untreated soil starts when the layer of soil aggregates is reaching saturation. This happens after 30–40 minutes for the 50 mm/hr rainfall and after 55–60 minutes for the 30 mm/hr rainfall. Before that, soil suction is able to absorb most of the rainfall.

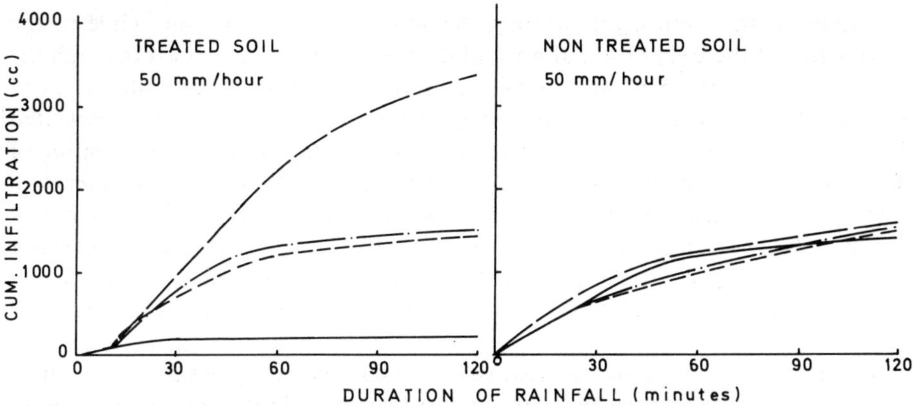

Figure 4.5. Cumulative infiltration of rainfall at 50 mm/hr into aggregates of the untreated and bitumen treated silt loam at zero slope gradient. Aggregate sizes: —— <2 mm; ---- 2–4·2 mm; –·–·– 4·2–5·3; ——— 5·3–8 mm

The volume of cumulative infiltration as a function of time for treated and non-treated soil with no slope, under a 50 mm/hr rainfall, is shown in Figure 4.5. Almost no infiltration of water occurs in the aggregates <2 mm of treated soil. The infiltration reaches similar values in 2–4·2 and 4·2–5·3 mm aggregates of both treated and non-treated soils, but much higher values for 5·3–8 mm aggregates of treated soil. All the infiltrated water in the non-treated soil

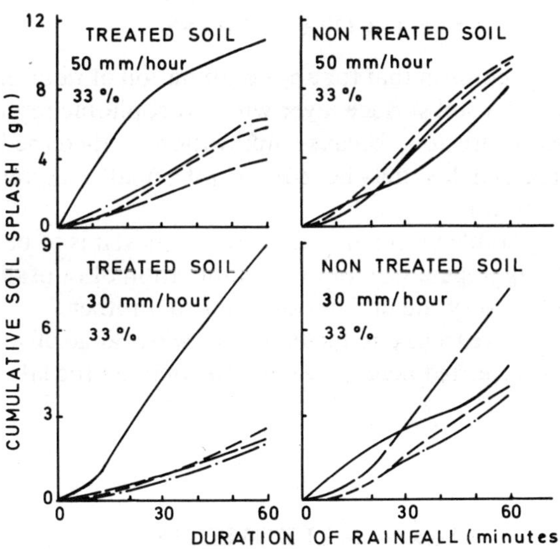

Figure 4.6. Cumulative soil splash from the untreated and bitumen treated silt loam at rainfalls of 50 and 30 mm/hr and a slope gradient of 33 per cent. Aggregate sizes:
—— <2 mm; --- 2–4·2 mm; –·–·– 4·2–5·3;
——— 5·3–8 mm

remained in the 5 cm layer, wetting the soil close to saturation, while in the aggregates of the treated soil almost all the infiltrated water passed through the pores between the aggregates, which remained dry. Only a few millimetres of slaked soil in the surface of the treated aggregates was found to have been wetted by the rainfall water at the end of the tests. For a 50 mm/hr rainfall, infiltration of water was reduced by hydrophobic treatment if the aggregates were smaller than 2 mm, and increased by the same treatment if the aggregates were larger than 5·3 mm. Even for aggregates 2–4·2 mm and 4·2–5·3 mm, the infiltrated water should be more effectively protected against evaporation in the treated soil, because it would be stored below the aggregated surface layer, which remains essentially dry.

Figure 4.6 presents some selected values of cumulative splashed soil, to show the correlation of soil splashed with soil loss by runoff, both for treated and non-treated soil, shown in Figures 4.3 and 4.4. Such correlation is very high for all aggregates in the non-treated soil and for aggregates <2 mm in the treated soil, but no correlation is observed in the case of larger aggregates of soil treated with hydrophobic bitumen emulsion. This suggests, as was observed by Gabriels *et al.* (1974), that the splashed material may be composed of stable aggregates, large enough not to be carried by runoff. The same kind of correlation was observed in the non-reported values for other rainfall intensities and slopes.

4.5.4. CONCLUSIONS

The results of the tests confirm that for any combination of rainfall intensity and slope, the treatment of a soil surface layer with hydrophobic bitumen emulsion may help to decrease erosion, because infiltration is increased, but only if aggregate size is large. Soil loss may be reduced with smaller aggregate sizes than required to prevent runoff.

Working with very stable bitumen-treated aggregates, it is to be expected that the critical values of aggregate size for any combination of rainfall intensity and slope would not differ very much from one soil to another.

More experiments have to be conducted with a wide range of soil types to test and refine the results reported here, especially to compare the laboratory results with field values.

ACKNOWLEDGMENTS

The experiments were carried out in the Department of Soil Physics of the State University of Ghent (Belgium) while the author was on leave from the Universidad Central de Venezuela.

The assistance provided by Prof. Dr. Ir. M. De Boodt, Dr. Ir. D. Grabriels and W. Minjauw is especially acknowledged.

REFERENCES

De Boodt, M., 1972, Improvement of soil structure by chemical means, In D. Hillel (Ed.), *Optimizing the Soil Physical Environment Toward Greater Crop Yields*, Academic Press, New York, pp. 43–54.

De Boodt, M., H. Verplandke and F. De Bisschop, 1974, Soil conditioning and its effect on water use, In *Isotope and Radiation Techniques in Soil Physics and Irrigation Studies* (1973), International Atomic Energy Agency, Vienna, pp. 417–433.

Gabriels, D., and M. De Boodt, 1972, Theoretical and practical approach for determining the optimal moisture content at the moment of soil conditioning, *Mitteilgn. Dtsch. Bodenkundl. Gesellsch.*, **15**, 185–203.

Gabriels, D., M. De Boodt and W. Minjauw, 1973, Description of a rainfall simulator for soil erosion studies, *Mededelingen Fakulteit Landbouwwetenschappen, State University Ghent*, **38** (2), 294–303.

Gabriels, D., R. Hartmann and M. De Boodt, 1974, Soil splash and runoff from untreated and chemically treated silt loam aggregates, *Mededelingen Fakulteit Landbouwwetenschappen, State Univ. Ghent*, **39** (3), 1471–1477.

Pla, I., 1975, Effects of bitumen emulsion and polyacrylamide on some physical properties of Venezuelan soils, In *Soil Conditioners*, pp. 35–46, Soil Sci. Soc. Amer., Spec. Pub., Madison, Wis.

Pla, I., G. Campero and R. Useche, 1974, Physical degradation of agricultural soils in the Western Plains of Venezuela, *Trans 10th Int. Congress of Soil Sci. Moscow*, **I**, 231–240.

4.6

Economics of Soil Conservation and Erosion Control Practices in Nigeria

S. O. OLAYIDE and A. O. FALUSI

4.6.1. INTRODUCTION

The state of agriculture in Nigeria is far from satisfactory. While other sectors of the Nigerian economy have recorded impressive growth rates in recent years, the average growth rate of the agricultural sector from 1970 to 1975 has been about 2·4 per cent. This cannot be said to be sufficient to keep pace with the ever-increasing population, and the increasing urban component of the total population with rising *per capita* incomes.

The situation in the food crop subsector is even more disturbing. A recent quantitative study of the food situation in Nigeria has portrayed Nigeria as a food deficit country (Olayide *et al.*, 1972). The study went on to warn that the food gap may be increasing unless urgent action is taken to modernize agricultural practices, particularly in the areas of food production.

One of the important factors responsible for the poor performance of the agricultural sector is the quantitative underutilization as well as the qualitative depreciation of land. Land, which is an important factor of production, is not only underutilized in a quantitative sense; the qualitative depreciation of most of the land under regular cultivation is even more apparent. An important cause of the qualitative depreciation of agricultural land is soil erosion.

4.6.2. EROSION PROBLEMS IN NIGERIA

4.6.2.1. Causes

In many areas where shifting cultivation is practised, whether in the forest zone or the savannah zone, the removal of the natural vegetative cover exposes the soil to the action of the elements, especially of rain or wind. The type of element causing erosion is mostly determined by the climate. In the forest zone, the main causative agent is rain, while in the dry savannah it is mainly wind. In the moist savannah, both agents are at work. Some of the natural factors which tend to accelerate soil erosion include soil type and topography (Ekern, 1950; Kowal, 1970b).

4.6.2.2. Extent of the Problem

Even though there are no firm statistics on the extent of soil erosion and fertility depletion in Nigeria, there is evidence to show that, in many parts of the country, the top soil is being increasingly ruined by soil erosion which is accentuated by outdated and unscientific farming practices.

Scattered information exists to show that in many areas of the forest zone, soil erosion has washed away between one-quarter to three-quarters of the top soil with consequent fertility depletion (Lal, 1973; Obi and Tuley, 1973; Datiri, 1973; Nye and Greenland, 1960; Grove, 1952). Erosion problems in some areas (particularly in the East Central State, Plates 1, 20 and 22) have advanced from sheet to gully erosion (Ofomata, 1964). Farmlands have also been reported carried away and completely ruined, especially in the Awka division of East Central State (Datiri, 1973).

Soil losses due to the action of the wind have also been reported in some parts of the Northern States of Nigeria (Chalk, 1963; Jones and Wild, 1975).

4.6.2.3. Effect of Soil Erosion

The effects of soil erosion on the physical properties of the soil, soil profile and nutrient status have been discussed previously (Ekem, 1950; Kowal, 1970a, b, 1972; Lal, 1976; Fapohunda, 1975). The inherent low-nutrient status of Nigerian soils, soil erosion losses and accelerated fertility depletion due to soil erosion and leaching, when coupled with the low-production technology used by our peasant farmers, are indicative factors of the low crop yields obtainable on most Nigerian farms. Present yields on most Nigerian peasant farms are between 10–20 per cent of potential levels under proper farm management (Olayide *et al.*, 1972).

4.6.3. METHODS OF COMBATING SOIL EROSION

Nigerian farmers have traditionally adopted the system of shifting cultivation to control erosion, weeds and pests and to maintain soil fertility for annual crops. The system, however, requires an extensive use of land. In the early days when shifting cultivation was practised, there was a balance between human population and soil fertility and therefore restricted soil erosion (Datiri, 1973). As long as the assumptions of a limitless supply of land and a low ratio of population to land hold good, the principles of soil conservation involved in the shifting cultivation system are as sound as any system can be under similar circumstances (Oluwasanmi, 1966; Ruthenberg, 1971). The increasing population pressure on land and the considerable pressure of food requirements have in recent years led to the extension of cropping areas as well as of cropping period; the latter has led to the shortening of fallow periods. Thus, there has been more land cultivated as well as a more intensive use of the land, and in the absence of sound soil conservation measures, soil erosion is likely to be aggravated, while soil fertility declines very rapidly (Nye and Greenland, 1960; FAO, 1966; Oyenuga, 1967; Anonymous, 1973).

There are, however, many modern soil conservation practices which can be used to combat the soil erosion problem in Nigeria. For purposes of exposition, we shall divide these practices into two classes. In the first category are those soil conservation practices which can be practised by farmers with little or no government assistance. These include mulching, contour farming, strip-cropping, tie-ridges, crop rotation, planting of cover crops, crop residues and minimum tillage practices.

In the second category are those soil conservation practices that, because of the high expense involved in their installation, can only be provided by the government or applied by farmers only if subsidized by the government. These include gully control, terracing, wind brakes, ponds and earth dams, irrigation dams, land smoothing and land grading, to mention just a few.

4.6.4. CONSTRAINTS ON PEASANT FARMERS IN THE ADOPTION OF SOIL CONSERVATION MEASURES

A number of constraints could hinder the adoption of sound conservation practices by peasant farmers. Notable among these are: (i) lack of awareness or conviction of the benefits of soil conservation; (ii) land tenure relationships; and (iii) lack of adequate working capital. Farmers may be unaware or be less than totally convinced of the benefits of soil conservation. When the former is the case, an expanded and reinvigorated extension education may be the solution to the problem. However, in the latter case, researchers would have to be prepared to provide packages of soil conservation measures which were technically feasible, economically profitable and socially acceptable to the farming community. For example, mulching, which is regarded as a good soil conservation measure, may not be very practical and economical, if the mulch material has to be grown on land that could have been used for economic crops, and if it has to be cut and transported, which is time-consuming and adds to the cost of growing cash crops.

The adoption of soil conservation measures is often an expensive venture. For this reason, when proper conservation measures are adopted in an area by a farmer it is unlikely that the farmer would want to leave the area for another, except, perhaps, when he needs to expand his farmland. Where a farmer does not have permanent title to a piece of farmland, he may feel very reluctant to adopt soil conservation practices. The present land tenure arrangements (particularly in the southern states of Nigeria), where land is communally rather than individually owned, may provide another disincentive for adopting soil conservation practices.

Conservation measures often require additional cash outlay, reorganization of farm enterprises, and long-term investments for which the returns may not be received for several years. Peasant farmers with limited capital and with serious pressure to expand current farming operations may not be able to afford some of the soil conservation and soil management practices.

Farmers need help in overcoming the multitude of obstacles to conservation measures. The questions to be answered are: what form should such help take, and how much of it is needed?

4.6.5. GOVERNMENT ACTION AND PUBLIC POLICY

Soil conservation as a goal of public policy is part of a superior end, that of husbanding the nation's natural resources in the interest of future generations. Since the time perspective of society is much longer than that of individuals, circumstances often make it profitable for the individual to exhaust or waste natural resources at such a rate that people coming several decades hence may suffer from want of them. It is, therefore, the duty of the government to protect the future productivity of land for generations to come.

Past government efforts in soil conservation programmes have been very meagre and have had little impact in solving the erosion problem in Nigeria. Governments have tended to treat the erosion problem as the private farmers' problem. The past two development plans in Nigeria have had little or no well-articulated soil conservation programmes. In the first National Development Plan document (1962–68), only the then Northern Regional Government had

Table 4.19. Federal Republic of Nigeria: Soil conservation programmes in the Third Development Plan 1975–80. Source: FMED, Federal Republic of Nigeria: *Third National Development Plan*, pp. 85–90.

Government	Capital allocation (₦ million)	Target area (ha)
Federal	1·0	N.a.
Benue-Plateau	3·0	N.a.
East Central	1·0	6070
Kano	5·0	9500
Kwara	0·5	N.a.
Lagos	—	—
Mid-Western	0·5	N.a.
North Central	0·5	N.a.
North Eastern	1·36	16,000
North Western	0·8	N.a.
Rivers	0·3	1000
South Eastern	1·0	N.a.
Western	0·4	80,000
	15·36	

N.a., not available.

capital allocation for soil conservation, and even then the plan had this to say: 'Soil conservation is an essential part of good farming and is thus the ultimate responsibility of the farmer himself' (Federal Government of Nigeria First National Development Plan, 1962, p. 115).

The Second National Development Plan (1970–74) was hardly any better from the point of view of soil conservation. Though increased emphasis was laid on irrigation schemes, a few soil conservation schemes were lumped together with the irrigation schemes and there were few or no guidelines as to the action programmes to be undertaken.

The recently launched Third National Development Plan (1975–80) is a welcome departure from previous plans. Bold and imaginative action programmes are proposed to be undertaken to conserve the soil resources of the country. Capital allocations for soil conservation programmes during the plan period are shown in Table 4.19.

Soil conservation programmes to be undertaken include contour bunding, terracing, check-draining, drainage systems, reclamation of large tracts of mining lands, planting of trees as shelter belts and the establishment of a soil and water conservation training centre.

4.6.5.1. The Need for Government Intervention

We have enumerated in an earlier section some of the constraints on the peasant farmers in the adoption of sound soil conservation practices. In this section we hope to highlight some of the reasons why government intervention is necessary.

4.6.5.2. Private and Social Cost

It is difficult to gauge the economic value of erosion losses and of conservation benefits. In fact, few studies exist which have attempted to measure the economic benefits of soil conservation practices. We can confidently say that both erosion losses and conservation benefits tend to be greater for society than for the individual because of the different time perspective of their respective plans. In the process of exploiting the soil, the farmer can often reap higher net incomes for a number of years than he could on a conservation basis. After such a period, and after the soil's productive capacity is seriously damaged, he may try to move on to another farm, thus leaving a worn-out piece of land that is of little use to anyone. The economic cost of erosion losses may not appear on his cost accounts, especially if he is a tenant and does not need to bear the cost of depreciation in land quality. To the community at large, however, this means less output for the same or higher input, and hence a clear-cut economic loss, which could have been avoided at a smaller cost than the loss.

This last proviso is the crux of the economic problem of measuring conservation benefits and costs. It inevitably involves two intricate determinations: (1) a comparison of the value of different goods in different points of time and of input costs at present and product returns in the long future; and (2) a segregation of present costs and future benefits between individuals and society. There may be other losses involved, besides the loss of future incomes, which also do not appear on the individual farmer's cost account such as silting of reservoirs and navigable streams, increased flood damage as a result of unretarded water runoff from the watershed, and dust-storm damage. It is because of these divergences between social and private benefits and costs that public policy and government involvements are essential in soil conservation programmes.

4.6.6. RESEARCH REQUIREMENTS

4.6.6.1. Design of Adoptable Farming Systems

There is a serious lack of knowledge of what the most effective combination of erosion control practices is under widely varying physical and economic conditions. The greatest challenge facing the agronomist and his economist counterpart is the design of farming systems that will permit continuous cultivation while being, at the same time, well adapted to the needs of the farming community. Sound soil conservation and soil management practices should be an integral part of such a farming system. Since soil and climate, topography, and farming practices vary widely over the country, adaptation to local conditions is essential. Specific practices should suit specific local conditions.

Though certain soil conservation measures may not be undertaken by the peasant farmer because of the initial high capital outlay, there are certain simple soil conservation practices which could be adopted by farmers. The growing of cover crops in mixed cropping practices, minimum tillage practices, fertilizer application, weed control, increased plant population density and improved cultural practices, described elsewhere in this volume, offer bright prospects. Researchers should, however, be cautious of introducing soil conservation practices whose technical feasibility, economic profitability and social acceptability have not been established.

4.6.6.2. Reinvigorated Extension Education

Assuming, of course, that packages of well-tested soil conservation and soil management practices exist for our farmers, there is the need for a reinvigorated extension education to make the benefits of soil conservation readily apparent to our farmers. The radio, newspapers, cinemas, posters and farmers' meetings are important avenues for getting the message across to the farmers.

4.6.6.3. Erosivity Index of Major Ecological Zones

As previously suggested (Chapter 3.1), there is an urgent need to have an erosivity index of the major ecological zones of Nigeria. This information is necessary in order to predict successfully the potential erosion hazard; it would also be helpful in planning farming operations and in following optimum conservation practices.

4.6.6.4. Erodibility Index of Nigerian Soils

Research should also be stepped up on the erodibility index of Nigerian soils. Preliminary research results on this aspect at IITA are very encouraging (Lal, 1976). The soils can be classified into different categories as regards their susceptibility to erosion, and appropriate soil conservation practices can be developed.

4.6.7. SUGGESTIONS FOR POLICY

4.6.7.1. Step Up Manpower Training and Development

There is a critical need for the training of personnel in practical aspects of soil conservation in the country. There is at present a dearth of well-trained soil scientists (soil physicists, soil chemists) and agricultural engineers, to mention just a few. Universities and agricultural schools should include training in soil conservation agronomy as part of their curriculum.

4.6.7.2. Increased Budget Allocation for Soil Conservation Programmes

Even though a substantial amount of money has been budgeted for soil conservation measures in the Third Development Plan period, it may not be sufficient if the country is to grapple seriously with the problems of qualitative depreciation of agricultural land. There is a need for government to increase the budget allocation for soil conservation programmes.

4.6.7.3. Establish Soil Conservation Districts/Zones

It is not too early for policy-makers, planners and administrators in this country to start thinking seriously about setting up soil conservation districts/zones particularly for agricultural land. Such a conservation programme for farmlands should have as its main objective the maintenance of soil productivity which, in turn, is a means to the superior end of getting maximum production from the nation's land resources in the long run.

Depending on the topography, soil type, rainfall intensity and/or wind velocity of an area, the government, acting on sound technical and professional advice, would encourage farmers in such an area to follow tested soil conservation practices. Such soil conservation zones would depend upon the voluntary cooperation of the farmers and on government assistance in regard to technical advice, professional personnel, equipment and supplies. Outright cash grants or input cost subsidies could be used to enlist farmers' cooperation.

4.6.8. CONCLUSIONS

There is an urgent need to modernize the agricultural sector of the Nigerian economy. To achieve this aim, a multipronged attack needs to be launched on the structural, institutional, socio-economic and cultural factors impeding progress on the agricultural front. Qualitative depreciation of agricultural land due to soil erosion problems has been one of these important factors.

Our researchers should develop adoptable farming systems that incorporate well-tested soil conservation and management practices. The technical feasibility, economic profitability and social acceptability of such systems should be uppermost in the minds of our researchers, planners and administrators.

Research should be stepped up into developing erosivity indices for the major ecological zones and erodibility indices of Nigerian soils.

Due to the expensive nature of some soil conservation measures, and because of the apparent divergence between the social and private costs and benefits of soil erosion and soil conservation, governments should become more active in soil conservation schemes than hitherto. Increased budget allocation, outright grants and/or subsidization of some costs of soil conservation borne by private farmers, stepping up of manpower, training and development in the field of soil conservation and the encouragement of the formation of soil conservation districts/zones are some of the ways the government can bring home to the people the need to husband our natural resources prudently for the benefit of present users and in the interest of future generations.

REFERENCES

Anonymous, 1973, Country report for Nigeria, Paper presented at the Regional Seminar On Shifting Cultivation and Soil Conservation In Africa, FAO/SIDA/ARCN, Ibadan.

Bennet, H. H., 1939, *Soil Conservation*, McGraw-Hill, New York.

Chalk, A. T., 1963, Soil conservation in Northern Nigeria and suggested programme, Consultant Rep. C-34, USAID.

Dabasi-Schweng, L., 1973, Economic aspects of shifting cultivation, In Shifting Cultivation and Soil Conservation In Africa, *FAO Soils Bull.*, No. 24, 78–98.

Datiri, B. T., 1973, Soil and water conservation practices, particularly in areas of shifting cultivation, In Shifting Cultivation and Soil Conservation In Africa, *FAO Soils Bull.*, No. 24, 237–241.

Ekern, P. C., 1950, Raindrop impact as the force initiating soil erosion, *Soil Science Soc. Amer. Proc.*, **15**, 7–10.

FAO, 1966, *Agricultural Development in Nigeria 1965–1980*, FAO, Rome.

Fapohunda, F. S., 1975, A quantitative analysis of runoff and infiltration as functions of cumulative rainstorm on a loamy soil, Students Project Report, Department of Agronomy, University of Ibadan.

Federal Government of Nigeria, 1962, *First National Development Plan (1962–1968)*, Federal Ministry Of Economic Development, Lagos, Nigeria.

Federal Government of Nigeria, 1970, *Second National Development Plan (1970–1974)*, Federal Ministry of Information, Lagos, Nigeria.

Federal Government of Nigeria, 1975, *Third National Development Plan (1975–1980)*, Federal Ministry of Economic Resources, Lagos, Nigeria.

Grove, A. T., 1952, Land use and soil conservation on the Jos plateau, *Nigeria Geological Survey Bulletin*, **22**, 63.

Jones, M. J. and A. Wild, 1975, *Soils of the West African Savanna*, Technical Comm. No. 55, Commonwealth Bureau of Soils, Harpden, Herts.

Kowal, J., 1970a, The hydrology of a small catchment basin at Samaru, Nigeria: III. An assessment of surface runoff water under varied land management and vegetative cover, *Nigerian Agricultural Journal*, **7**, 120–133.

Kowal, J., 1970b, The hydrology of a small catchment basin at Samaru, Nigeria: IV. Assessment of soil erosion under varied land management and vegetative cover, *Nigerian Agricultural Journal*, **7**, 134–147.

Kowal, J., 1972, Soil–plant–water relationships at Samaru, Zaria, Tropical Soils Research Seminar, IITA, Ibadan, Nigeria.

Lal, R., 1976, *Soil erosion investigations on an Alfisol in southern Nigeria*, IITA Monograph No. 1.

Mouttapa, F., 1973, Soil aspects of shifting cultivation, In Shifting Cultivation and Soil Conservation In Africa, *FAO*, *Soils Bull.*, No. 24, 37–47.

Nye, P. H., and D. J. Greenland, 1960, *The soil under shifting cultivation*, Technical Comm. No. 51, Commonwealth Bureau of Soils, Harpenden, Herts.

Obi, J. K., and P. Tuley, 1973, The bush fallow and ley farming on the oil palm belt of Southeastern Nigeria, Unpublished Misc. Report, Land Resources Division, Overseas Development Administration No. 161.

Ofomata, G. E. K., 1964, Soil erosion in the Enugu Region of Nigeria, *African Soils*, **9**, 289–320.

Olayide, S. O., D. Olatunbosun, E. O. Idusogie and J. D. Abiagom, 1972, *A Quantitative Analysis of Food Requirements, Supplies, and Demands in Nigeria 1968–1985*, Federal Department Of Agriculture, Lagos, Nigeria.

Oluwasanmi, H. A., 1966, *Agriculture and Nigerian Economic Development*, Oxford University Press, Ibadan.

Oyenuga, V. A., 1967, *Agriculture in Nigeria*, FAO, Rome.

Piersen, G. L., 1973, *Agronomic Aspects of Soil Conservation. In Shifting Cultivation and Soil Conservation in Africa, FAO. Soils Bull.*, No. 24, 235–236.

Ruthenberg, H., 1971, *Farming Systems in the Tropics*, Oxford University Press, London.

4.7

Evaluating Systems of Soil Conservation through Bioeconomic Modelling

R. G. DUMSDAY and J. C. FLINN

4.7.1. INTRODUCTION

The world population in the past 100 years has increased from some 1·2 billion to approximately 3·9 billion people. The result has been that the effective area of land available per person for agriculture and animal production has been reduced by two-thirds over the period, with a consequent intensification of agricultural production relative to the land resource. As the pressure on land within the humid tropics increases, the duration of the bush fallow phase of the farming system decreases, accompanied by a decline in soil fertility and in the productivity of both land and labour (Grove, 1951; Lagemann *et al.*, 1976). Thus, to assess the long-term technical and economic implications of alternative management strategies to increase food-crop production in these zones requires that predictive models of farming systems be developed which take explicit account of the interactions between soil fertility, erosion, weather and agricultural production.

Farming systems in the tropics, as elsewhere, are a consequence of the interaction of complex biophysical and socioeconomic relationships. Further, the time sequence of the flow of inputs, many of which are beyond the control of the farmer, and the flow of outputs results in the farmer operating in a dynamic, uncertain environment. It is within such a conceptual framework that the evaluation of soil conservation and associated management strategies in agriculture should ideally take place. However, whether the evaluation of alternative soil management strategies does take place within such a framework is influenced by a number of considerations. Among the more important are:

(a) the current state of knowledge concerning the relationships of the system;
(b) the cost, time involved, and technical feasibility of gathering additional information; and
(c) the techniques available for analysing the information.

Past attempts to evaluate soil conservation strategies at the farm level have been limited for a number of reasons. First, the analytical models used were deterministic, and no provision was made for the variable nature of the erosion process, with its resultant influence on costs and returns and thus on patterns of farm income through time. Second, the data used were usually highly specific, for example, with respect to management strategies, soil types and weather conditions, which limited the generality of the conclusions drawn from the studies. Third, the models represented marked abstractions from reality, due in part to a lack of knowledge concerning system interrelationships, but also due to computational limitations of the techniques used.

One means of overcoming some of the above restrictions is through the use of simulation techniques, which are ideally suited to the analysis of complex systems (Dent and Anderson, 1971). In a simulation study, a model is constructed which is anticipated to behave in a manner similar to the system in the real world. The 'performance' of the system under alternate input patterns and decision rules may then be studied in an experimental environment without actually incurring the penalties of putting the system into operation. There are, of course, many difficulties associated with both model construction and subsequent experimentation. These problems include: identifying the important variables; defining the structure of the model; the availability of data for specifying relationships in the model; validation of the model and its components; and the interpretation of results obtained from experimentation with the model.

In this chapter, a simulation model designed to evaluate various soil conservation practices is discussed. The study area for which the model was built is an intensively cropped region with a relative high potential for soil erosion in the subtropical wheat-growing areas of east-central Australia. In many ways, the topography is similar to large areas of the derived savannah zones of West Africa, where sheet, as opposed to gully erosion, is the major soil conservation problem and where extensive tracts of land are being cleared by government and private individuals for large-scale, mechanized farming.

4.7.2. THE BIOECONOMIC MODEL

Figure 4.7, modified from Dumsday (1971), depicts the more important relationships included in the bioeconomic simulation model of the soil conservation system. The circled factors in Figure 4.7 represent the points of entry into and exit from the system. For example, specification of soil type determines the internal factors of inherent soil erodibility, infiltration characteristics and depth of soil horizons. The choice of cropping system implies a possible set of cultivation practices, sowing and harvest dates. Thus, the factors involving input–output prices, production costs, mechanical control practices, topography, soil type, cropping system and weather set the boundaries of the system to be synthesized. The values adopted for these factors, in association with the assumed biophysical relationships between the internal factors, ultimately determine the output from the system of prime interest, in this case, net revenue.

Space does not permit a detailed description of the model, which has been reported by Dumsday (1971, 1973). Suffice it to mention that the model was

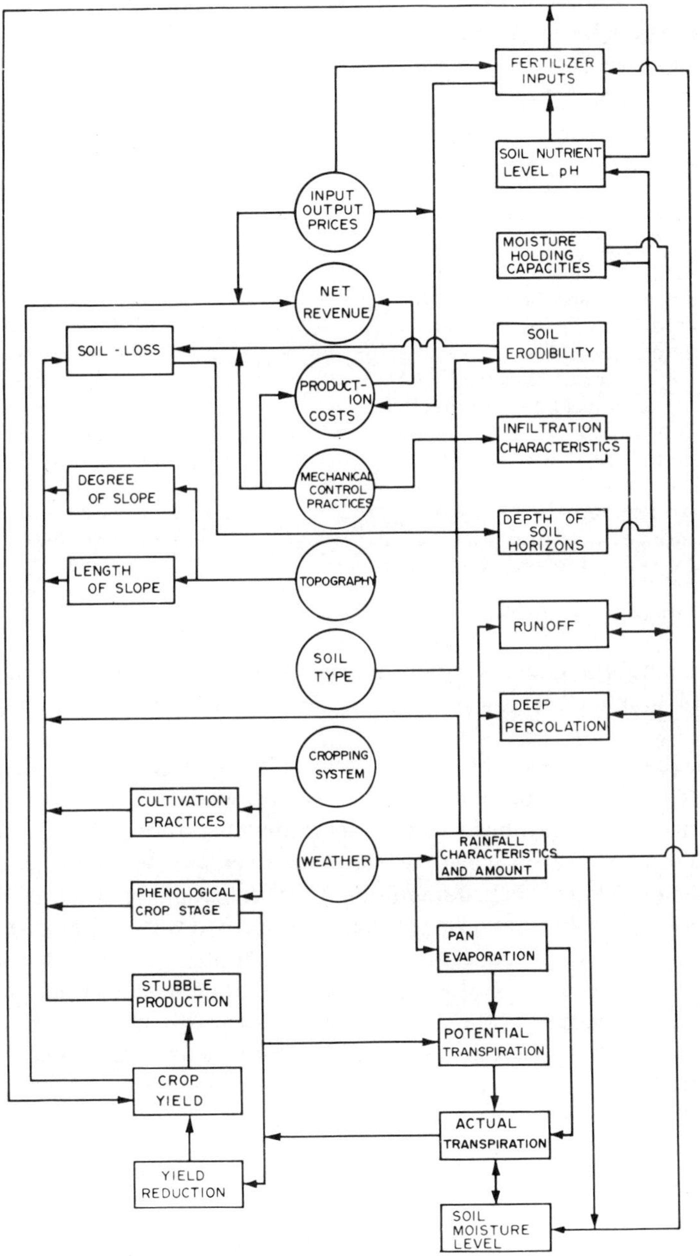

Figure 4.7. Interrelationships of the soil conservation system

based largely on data and relationships reported by the University of New England (1968), Boughton (1965), Fawcett (1967), Wischmeier and Smith (1958, 1962, 1965), the New South Wales and Queensland Departments of Soil Conservation, and the Australian Bureau of Meteorology. The empirical forms of these relationships and their uses and limitations are discussed by Dumsday (1973). The more important relationships used in this model, and their sources, are listed in Table 4.20.

Table 4.20. Relationships embedded in the simulation model

Relationship	Source
Weather sequences	From site and generated
Kinetic energy of rainstorm, *KE*	W and S* (1958)
Index of storm erosion potential, EI_{30}	Derived from met. data
Inherent erodibility of soil type, *K*	W and S (1962)
Effect of slope on soil loss, *LS*	W and S (1965)
Cropping management factors, *C*	W and S (1965)
Factors for mechanical control practices, *P*	W and S (1965)
Estimation of runoff	Boughton (1965)
Stubble production	Fawcett (1967)
Soil-nutrient profiles	Fawcett (1967)
Response to fertilizers	Dumsday (1973)
Soil moisture characteristics	Fawcett (1967)
Soil moisture–crop growth relationships	Flinn (1968)

* W and S, Wischmeier and Smith.

The simulation model essentially consisted of two separate sections. The first dealt with biophysical aspects and the second with economic aspects of the system.

The biophysical section of the model was divided into three phases. The first phase used initial data concerning the biophysical environment, i.e. soil characteristics, topographical factors, mechanical relationships, soil-loss ratios as related to stage of crop growth and stubble cover, and potential crop yields for the specified soil type. These data initiated the second phase of the model, which operated on a daily basis and was primarily concerned with computing, given the present state of the system, crop growth and soil loss as a consequence of the weather which prevailed on that day. The weather data were either drawn from an historical sequence or generated from historical distributions. The remaining phase of the biophysical section accumulated soil loss for each production period and updated the depth of the surface soil horizon to its value for the subsequent production period. The output of the biophysical section of the model consisted of annual totals for climatic variables (evaporation, erosive potential of storm rain, rainfall) and system outputs (deep percolation, runoff, soil loss and crop yields).

The economic section of the model was initiated by setting prices for fertilizer inputs and crop outputs, costs of construction, maintenance and reconstruction associated with mechanical control practices, and those variable crop production costs which were assumed constant for a given simulation run. Parameter values were then specified for the fertility gradients of the soil profile, fertilizer response, proportion of land removed from production by mechanical control practices, and discount rates. The biological and management components of the model interacted to generate actual crop production over a series of environments and management strategies. Prices and costs were used to weight the values of outputs from and inputs to the system over time, resulting in a discounted net revenue stream being developed for each strategy. For each production period of the simulation run the model outputs were fertilizer allocation and discounted net revenue per hectare for the specified discount rates.

4.7.3. SOIL MANAGEMENT PRACTICES

A farmer is faced with a large array of alternatives for manipulating those factors in his farming system which are under his control. The manager's problem is to select, within his constraint set, those policies or decision rules which, when implemented, assist him in moving towards his personal goals. When considering soil management, there are three groups of practices, discussed below, which will importantly determine the rate of soil loss, crop yields and, as a result, farm profits over time.

One managerial choice faced by the farmer is that of the length and type of summer fallow. Because much of the annual rainfall in the study area of Australia is received in the summer, most farmers maintain their land under a bare fallow at this time for four to six months. An increase in the length of bare fallow may, in many seasons, result in an increase in the quantity of moisture stored for use by the subsequent crop. However, there will also usually be a concomitant increase in the frequency of cultivations and in the exposure of soil to erosive storms. For a given length of bare fallow the frequency and type of cultivation may be varied, or intra-fallow cultivation may be replaced by chemical weed control.

A second set of decisions concerns the treatment of crop residues. Crop stubble may be mulched or burned. The effectiveness of stubble mulching in reducing erosion during fallow and the early stages of growth of the subsequent crop will largely depend on the quantity of stubble, which in turn varies with the production of vegetative matter by the previous crop. Disadvantages of stubble mulching may be an increased requirement for applied nitrogen and mechanical difficulties in cultivating where large quantities of stubble are present. In the study area, approximately 50 per cent of farmers burnt their stubbles while the other 50 per cent practised stubble mulching.

The third set of decisions concerns the use of mechanical control measures. These measures included contour cultivation, contour strip-cropping, and contour and graded terrace (or bank) systems. Terrace systems divide slopes into discrete segments, effectively reducing slope lengths. Various terrace types, spacings and capacities will result in different levels of soil and moisture conservation and different costs.

In addition to the three sets of management decisions discussed above, the model takes specific account of the effect of soil loss on the optimal rates of fertilizer application. That is, the trade-off between substituting fertilizer for eroded top soil can be assessed within the model.

4.7.4. DESIGN OF SIMULATION EXPERIMENTS

The model can be used to run numerous experiments to compare large numbers of treatments and weather sequences. However, in common with most research projects, the objectives of the researchers, and available time and funds, imposed limits on the amount of experimentation actually performed.

Each of the treatments selected for study were run over five separate 50-year periods using statistically generated weather data. The choice of a 50-year period was largely arbitrary, but, a number of considerations led to the use of this time horizon. First, experience when developing the model had shown that the differences between most treatments had become clearly established within a period of 20 to 50 years. Secondly, farmers in the study area seemed to regard 50 to 70 years as an upper limit to their planning horizons. Thirdly, and possibly most important from a planning viewpoint, uncertainty concerning future technological and economic change increases with time and has the effect of reducing the usefulness of information generated for distant time periods. Finally, for the non-zero discount rates chosen the relative weights assigned to net revenues became small by the time the fifth decade was reached. (Discount rates were chosen to represent a zero rate of time preference (0 per cent), and opportunity rates of return on low-risk investment (6 per cent), medium-risk investment (10 per cent), and marginal investment (25 per cent). The zero discount rate represents conservation in the strict sense.)

Table 4.21. Experimental environments used in simulation runs

Environment number	Soil type*	Depth of horizon A (mm)	Depth of horizon B (mm)	Length of field slope (m)	Degree of field slope (%)
1	RBE	50	150	60	4
2	RBE	50	150	180	12
3	RBE	100	300	120	8
4	RBE	150	450	60	4
5	RBE	150	450	180	12
6	BE	100	300	120	2
7	BE	100	300	360	6
8	BE	150	450	240	4
9	BE	200	600	120	2
10	BE	200	600	360	6

* RBE, red-brown earth (Ultisol); BE, black earth (Vertisol).

Table 4.22. Management systems evaluated in simulation runs

Management system	Management practices	Key to management practices
1	a_1 b_2 c_2 d_3	a stubble management
2	a_2 b_2 c_2 d_3	a_1 stubble left
3	a_1 b_1 c_2 d_3	a_2 stubble removed
4	a_1 b_3 c_2 d_3	
5	a_1 b_2 c_1 d_3	b length of bare fallow
6	a_1 b_2 c_3 d_3	b_1 2 months
7	a_1 b_2 c_2 d_1	b_2 4 months
8	a_1 b_2 c_2 d_2	b_3 6 months
9	a_2 b_3 c_3 d_3	
10	a_1 b_1 c_1 d_1	c frequency of cultivation
11	a_2 b_2 c_2 d_2	during fallow
		c_1 1 cultivation
		c_2 3 cultivations
		c_3 5 cultivations
		d mechanical control practices
		d_1 terracing
		d_2 contour cultivation
		d_3 up and down slope (vertical) cultivation

Ten 'environments' (Table 4.21) and 11 management systems (Table 4.22) were selected to give 110 experimental treatments. The choice of environments covered extremes from low levels of erosion hazard (environments 4, 9) to high levels of erosion hazard (environments 2, 7) and was centred on modal combinations of soil depth and field slope for each soil type. The choice of management systems covered extremes from low (system 10) to high (system 9) potential rates of use of soil resources and was centred on management practices most likely to be relevant to existing or potential cropping systems for the study area.

4.7.5. BEST MANAGEMENT SYSTEMS FOR EACH ENVIRONMENT

The most profitable management system of those tested in each environment, at each discount rate, was selected on the basis of the mean and coefficient of variation of the cumulative discounted net revenue generated by the simulation model for a 50-year period. (As the coefficient of variation gives a measure of the relative dispersion of outcomes about the mean, it was used as an indicator of the level of risk associated with a particular management system. Decision-making criteria which take explicit account of risk have been described (e.g. Dillon, 1971; Dillon and Anderson, 1971) and could be employed in analysis of the output generated by the model.) The management systems selected on this basis are presented in Tables 4.23 and 4.24 along with the mean soil loss and cumulative net revenue corresponding to each system. In general, systems incurring high rates of soil loss also suffered low mean levels of cumulative revenue and high coefficients of variation.

Table. 4.23. Break-even costs for minimizing soil loss on red-brown earths

Environment	Minimum soil loss (mm)	Discount rate (%)	Management system selected	Soil loss* (mm)	Cumulative net revenue ($/ha)	Break-even costs for minimizing soil loss ($/ha)	Adjusted break-even cost ($/ha)
1	14	0	10	14	693	0	0
		6	8	21	214	6	4
		10	8	21	139	12	12
		25	8	21	60	16	16
2	57	0	10	57	413	0	0
		6	10	57	127	0	0
		10	3	191	79	4	3
		25	3	191	49	32	32
3	29	0	3	101	1285	30	0
		6	3	101	452	67	64
		10	3	101	301	64	64
		25	3	101	135	50	50
4	13	0	10	13	2221	0	0
		6	8	14	718	12	11
		10	8	14	460	15	15
		25	8	14	193	16	16
5	50	0	10	50	1802	0	0
		6	3	250	612	41	32
		10	3	250	414	58	57
		25	3	250	186	54	54

* 1 mm = 16 tonnes/ha, approximately.

It was apparent from the simulation output that for all environments, from a long-term viewpoint, management systems combining stubble burning with vertical cultivation (i.e. systems 2 and 9) produced marked reductions in net revenue and increases in soil loss relative to the results for other systems. These effects became apparent after relatively short (5 to 10 years) periods of time. It was assumed that contour cultivation alone was not effective in reducing soil loss for some combinations of slope length and degree of slope. Thus, for environments 2, 3, 5, 7, 8 and 10, the net revenue streams for management systems 8 and 11 were identical to those for systems 1 and 2 respectively.

A broad conclusion that may be reached from the results shown in Tables 4.23 and 4.24 is that the best systems for all environments in the long term tend to be

Table 4.24. Break-even costs for minimizing soil loss on black earths

Environ-ment	Minimum soil loss (mm)	Discount rate (%)	Manage-ment system selected	Soil loss* (mm)	Cumulative net revenue ($/ha)	Break-even costs for minimizing soil loss ($/ha)	Adjusted break-even cost ($/ha)
6	10	0	10	10	2216	0	0
		6	8	15	709	5	3
		10	3	24	455	10	10
		25	3	24	192	13	13
7	18	0	10	18	1847	0	0
		6	3	136	629	54	46
		10	3	136	420	64	63
		25	3	136	186	55	55
8	14	0	10	14	2888	0	0
		6	5	54	917	2	0
		10	5	54	587	14	13
		25	5	54	242	18	18
9	10	0	8	13	3246	45	0
		6	8	13	1040	26	26
		10	8	13	658	23	23
		25	8	13	267	17	17
10	17	0	1, 8	122	3037	309	∞
		6	1, 8	122	1004	153	153
		10	1, 8	122	643	118	118
		25	1, 8	122	265	71	71

* 1 mm = 12 tonnes/ha, approximately.

those employing relatively low rates of use of soil resources. This conclusion held even for relatively high discount rates. However, the best system was not always that which minimized soil loss (system 10), even for a zero discount rate. The break-even costs for minimizing soil loss, shown in Tables 4.23 and 4.24, are the cumulative net revenue, at each discount rate, for the selected system less that for system 10.

The adjusted break-even costs in Tables 4.23 and 4.24 take account of the productive state of the soil resources at the end of the planning period. Thus, the standard formula for net present value was modified to take this fact into account. This adjustment resulted in management system 10 becoming the best system for environments 3 and 9 at zero discount rates. Other than that, the adjustment had little effect on break-even costs for minimizing soil loss.

Although an infinite adjusted break-even cost was associated with environment 10 at zero discount rate, it could be expected that the difference between the net revenue streams for systems 1, 8 and 10 would not remain constant in perpetuity, and that system 10 would eventually prevail.

As would be anticipated, management systems employing low (or near enough to zero) rates of use of soil resources were the best systems for all environments if production was considered in perpetuity and a zero rate of time preference was assumed. However, while these conditions meet the requirements of the 'true' conservationist, many farm operators, and some governments, may, for quite legitimate reasons, adopt non-zero discount rates.

For the non-zero rates assumed, the use of terrace systems (7, 10) is questionable for most situations. On the other hand, the best systems involved shorter periods of summer fallow than are commonly used in the study area. Chemical cultivation appeared in the selected systems for environment 8 (black earths), otherwise the best systems for red-brown earths and black earths were similar.

The benefits, in terms of moisture conservation, of frequent cultivation for weed control during the fallow period (system 6) appeared to be outweighed by the costs incurred through loss of soil structure. Contour cultivation, in the absence of other mechanical control practices, appeared in the best management systems for several environments.

Often farmers, as private decision-makers, have shorter planning horizons or employ higher discount rates than does society as reflected in government policies. These situations result in different 'intensities' of conservation being appropriate for the two groups, that is, the society at large would often prefer a more effective soil-conserving strategy than would the individual farmer. (For example, compare management policies which are optimal for the 0 and 25 per cent discount rates in Tables 4.23 and 4.24.) Governments have evolved numerous measures for such situations. They include provision of financial or technical grants; loan funds; taxation concessions; and compulsory legislation. Break-even cost data, as presented in Tables 4.23 and 4.24, provide useful guides to the nature and extent of the government action necessary to persuade individual farmers to meet the conservation objectives of society.

4.7.6. SOIL CONSERVATION MODELS FOR THE HUMID TROPICS

The model discussed above was developed for a subtropical semi-arid region characterized by large (approximately 400 hectare), highly mechanized farms where wheat is grown as a monoculture, and where moisture conservation and weed control by bare fallowing is important. In contrast, much of the agriculture in the humid tropics is characterized by small (less than 2 hectare), labour-intensive farms employing various systems of mixed cropping combined with shifting cultivation, and where bare fallowing is of course a totally unnecessary and unacceptable practice. The types of erosion control appropriate to larger farms are often completely irrelevant for the small farmer, who is likely to continue mixed cropping and shifting cultivation for several decades. Conservation systems for the smallholder will probably need to rely on the selection of crop, mulching, weed control and cultivation practices, and the timing of these operations for their success, as opposed to the use of mechanically constructed erosion control measures. The need to cater for a large number of crops, cropping systems, forms of cultivation and crop residue management will undoubtedly make the required bioeconomic models more complex than the one reported, in addition to making it costly and time-consuming to assemble the empirical data.

When developing a crop–soil erosion simulator for tropical systems, two assumptions, not biophysical in nature, reasonable in the subtropical model are questionable for the small-farmer system. First, in the model reported, the present value of the net income stream was the choice criterion for selecting the preferred soil management practices. The nature of the small farmer's preferences may suggest an alternative choice criterion (e.g. that which ensures stability of food production over time, minimizes labour or capital expenditure subject to a threshold level of output, etc.) as more appropriate. Second, the nature of the constraints faced by the small farmer—those of liquidity, scarcity of labour and often land—suggests that the nature of the erosion technology tested should be neither capital- nor labour-intensive. This fact may be particularly important, as it appears that many small farmers do not regard erosion as one of their major farming problems (Williams *et al.*, 1975).

Although agricultural production in the humid tropics is based predominantly on small-farmer systems, there are large areas under mechanized, sole-crop farming, particularly in the derived savannah zones. However, lack of erosion control has emerged as one of the most severe limitations of the technology employed (Baldwin, 1957). The model discussed could be readily adapted to predict the consequences of alternative strategies of land use on these large-scale farms, the main problem in using the model being the paucity of relevant soil, crop, weather and economic data.

Yet these data constraints are becoming less of a bottleneck. The work of Roose (1967) and Lal (1976), among others, which relates the erosivity of a storm to crop and soil conditions in West Africa, is providing the class of information required for development and use of crop-erosion models. Further, advances made in statistical generation of weather data from limited weather records (e.g. Phillips, 1971; Dumsday, 1973) may result in the lack of long-term weather records not being an impediment to the computer testing of various soil management strategies. Finally, it is worth noting that Linsley (1975) has been able to apply the model, as described in this chapter, to problems of maize production in Malawi.

SUMMARY

Both conceptual and data problems exist when developing simulation models of soil conservation systems for the humid tropics. However, the time is ripe for soil scientists, agronomists and economists to pool their expertise to develop such models. Simulation models may be constructed from data collected from a variety of sources, and hence progress can be rapid; this feature represents one of the chief advantages of simulation methods (Angus *et al.*, 1974), especially for regions which do not have a long history of agricultural research—a feature of many of the food crop producing areas in the humid tropics. Often these are the same regions where the pressures on agricultural land are greatest, where the productivity of land and labour is declining (Lagemann *et al.*, 1976), and where the threat of erosion is greatest. In such regions the continual demand for more productive and profitable crop production systems can make life very difficult for agricultural researchers relying on field and laboratory methods. The increasing availability of computer systems should enable these researchers to complement more traditional research methods with simulation procedures. This approach offers the possibilities of:

(a) increasing the number of new cropping systems that may be examined for a given bundle of research resources;

(b) reducing the time taken to obtain information on system performance under varying economic and weather conditions;

(c) markedly reducing the time and research resources required to examine the long-run bioeconomic effects of systems aimed at soil conservation; and

(d) providing the basis for more purposeful field and laboratory research.

REFERENCES

Angus, J. F., J. J. Basinski and H. A. Nix, 1974, Weather analysis and its application to production strategy in areas of climatic instability, *International Expert Consultation on the Use of Improved Technology for Food Production in Rain Fed Areas of Tropical Asia*, FAO, Rome.

Baldwin, K. D. S., 1957, *The Niger Agricultural Development Project*, An Experiment in African Development, Blackwells, Oxford.

Boughton, W. C., 1965, *A New Simulation Technique for Estimating Catchment Yield*, Water Research Laboratory, University of New South Wales, Report No. 78.

Dent, J. B., and J. R. Anderson (Eds.), 1971, *Systems Analysis in Agricultural Management*, Wiley, New York.

Dillon, J. L., 1971, Interpreting systems simulation output for managerial decision-making, In J. B. Dent and J. R. Anderson, (Eds.), 1971, *Systems Analysis in Agricultural Management*, Wiley, New York.

Dillon, J. L., and J. R. Anderson, 1971, Allocative efficiency, traditional agriculture and risk, *American Journal of Agricultural Economics*, **53**, 26–32.

Dumsday, R. G., 1971, Evaluation of soil conservation policies by systems analysis, In J. B. Dent and J. R. Anderson, (Eds.), 1971, *Systems Analysis in Agricultural Management*, Wiley, New York.

Dumsday, R. G., 1973, *The Economics of Some Soil Conservation Practices in the Wheat Belt of Northern New South Wales and Southern Queensland—A Modelling Approach*, University of New England, Farm Management Bulletin No. 19.

Fawcett, R. G., 1967, *Factors Influencing the Development and Grain Yield of Spring Wheats Grown on the North-West Plains of New South Wales, with Special Reference to Soil–Water–Plant Measurements*, Unpublished Ph.D. Thesis, University of Sydney.

Flinn, J. C., 1968, *Allocation of Water Resources. A Derived Demand Function for Irrigation Water, and Efficiency of Allocation*, University of New England, Farm Management Bulletin No. 1.

Grove, A. T., 1951, Land use and soil conservation in parts of Onitsha and Owerri provinces, *Geographical Survey of Nigeria*, Bulletin No. 21.

Lagemann, J., J. C. Flinn and H. Ruthenberg, 1976, Land use, soil fertility and agricultural productivity as influenced by population density in Eastern Nigeria, IITA, Mimeo, Feb. 1976.

Lal, R., 1976, *Soil erosion investigations on an Alfisol in Southern Nigeria*, IITA Monograph No. 1.

Linsley, J. L., 1975, *Land Husbandry for Greater Yields*, Unpublished M. Sc. Thesis, Department of Agriculture, University of Reading.

Phillips, J. B., 1971, Statistical methods in systems analysis, In J. B. Dent and J. R. Anderson, (Eds.), 1971, *Systems Analysis in Agricultural Management*, Wiley, New York.

Roose, E. J., 1967, Ten years of erosion and run-off records in Senegal, *L'Agronomie Tropicale*, **22**, 123–152.

University of New England, 1968, *Report to Australian Fertilizers Limited on the analysis of their N–P trials on wheat, 1964–67*, Farm Management Service Centre, University of New England.

Williams, A. V., J. C. Flinn and C. G. Knight, 1975, Farming problems as perceived by two groups of smallholders in South Eastern Nigeria, IITA, Mimeo, 1975.

Wischmeier, W. H., and D. D. Smith, 1958, Rainfall energy and its relationships to soil loss, *Transactions of the American Geophysical Union*, **39**, 285–291.

Wischmeier, W. H., and D. D. Smith, 1962, Soil loss estimation as a tool in soil and water management planning, *Commission on Land Erosion*, Publication No. 59, International Association of Scientific Hydrology, pp. 148–159.

Wischmeier, W. H., and D. D. Smith, 1965, *Predicting rainfall-erosion losses from cropland*, Agricultural Research Service, USDA, Agric. Handbook No. 282.

PART 5

Farming Systems and Erosion Hazard in Africa

5.1

The Soil and Water Conservation Problem in Africa: Ecological Differences and Management Problems

R. LAL

The forest region in Africa is continuously giving way to derived savannah, because of the increasing demand for food production. Roche (1973) reported that the forested region had decreased by 100 million hectares. Before discussing the differences in the erosion problems and management principles in the two environments, a brief review of erosion research and surveys is given. An earlier survey of the erosion hazard in Africa has been reported by Stockdale (1937) and a recent review has been produced by Lal (1976b).

5.1.1. SAVANNAH REGION

5.1.1.1. East Africa

Some of the earlier reports of erosion in East Africa have been by Thompson (1939), King (1939), and Ducker and Hoyle (1948). Recently a comprehensive review of erosion in Tanzania has been published by Rapp *et al.* (1972), and results of erosion research from Kenya have been reported by Gethin-Jones (1936), Maher (1950) and Brook (1955). Brook (1955) found that a 51-mm storm in 30 minutes caused a loss of nearly 3500 m^3 from a 10-hectare coffee field with a slope of 8°. Considerable work on the use of mulching techniques and herbicides for erosion control in coffee has been reported by Wallis (1961).

Erosion research data in Tanzania are available from Mpwapwa (6° 20′S, 36° 30′E), Tengeru (3° 22′S, 36° 48′E) and Shinyana (3° 35′S, 33° 25′E). Rapp *et al.* (1972) reported a negligible soil loss from thicket and grass plots. Wrigley (1969) reported erosion and runoff data from a catchment study at Mbeya Range. The maximum runoff from cultivated catchment was 76 cusec/m^2 as compared with 12 cusec/m^2 from the forested region.

Hutchinson *et al.* (1958) found from their investigations at Namulonge (0° 30′N, 32° 37′E) that 10 times more runoff occurred from bare plots than from grass-covered plots, and that grass mulch was twice as effective as stone mulch in terms of runoff control. Sperow and Keefer (1975) reported soil eriosion losses from the same research station in Uganda under different crops (Table 5.1).

Table 5.1. Effect of different vegetal cover on runoff and soil losses at Namulonge, Uganda (Sperow and Keefer, 1975)

Crop	Runoff (% of precip.)	Relative soil loss (tonnes/season)
No crop (bare plot, cultivated)	40·6	81·5
Maize, annually	23·4	34·0
Cowpeas, annually	19·9	26·6
Rhodes grass (permanent)	5·3	26·6
Veld, grazed and burned	10·8	2·5
Veld, closely grazed	5·6	4·4
Veld, intact	0·5	0·1

Extensive erosion work from Rhodesia has been reported by Hudson (1957a), Hudson and Jackson (1959) and Elwell and Stocking (1973a and b). Hudson (1957a) reported that annual soil loss from continuous maize plots over a period of six years was 6·7, 3·7 and 3·1 m^3/ha, respectively, from slopes of 3·5, 2·5 and 1·5°. The average annual soil loss from a bare fallow plot was 141·3 m^3/ha/annum as compared with 1·2 m^3/ha/annum from a weed-free bare plot with a wire gauge suspended above it. In addition to Hudson's $KE > 1$ index (Chapter 3.1) Elwell and Stocking (1973b, 1976) have reported that seasonal momentum values of all storms proved to be the best prediction of direct surface runoff yield.

5.1.1.2. West Africa

Detailed reviews of erosion research in the savannah region of West Africa have been published by Fournier (1967) and Jones and Wild (1975).

Available evidence indicates that though gullying is the more spectacular form of erosion in the savannah regions of Nigeria, sheet erosion is very serious, causing severe losses of top soil and applied nutrients. Surveys of gulley erosion from eastern Nigeria have been reported by Grove (1949, 1951), Ofomata (1964, 1965) and Floyd (1965). In northern Nigeria, Chalk (1963) estimated 20 to 50 per cent of the upland soils in the Chalwa catchment to be seriously gullied. Another report by FAO (1965) indicated that as much as 70 per cent of the soils of northern Nigeria had been eroded to some extent. On the Jos Plateau, Jones (1975) reported the extent of gully erosion caused by tin mining and uncontrolled grazing. It was estimated that the total length of active gullying in the Jos Plateau is about 4500 miles, representing approximately 100,000 tons of soil lost. The cost of mechanical repairs alone is estimated to be more than ₦30 million. Kowal

(1970b) reported runoff and soil loss data from Samaru, Zaria. Ridges down slope that had the highest runoff (28 per cent) and soil loss (20 tonnes/ha/annum). The least runoff loss was from alternate tied ridges (13 per cent) with a soil loss of about 6 tonnes/ha/annum.

In the savannah region of Ghana, reports indicate that gulley erosion is not as serious as sheet erosion (Brammer, 1962; Adu, 1972; Obeng, 1972).

Erosion in Francophone West Africa has been extensively reported by Charreau (1972), Roose (1967) and Charreau (1968). Charreau (1972) reported that the erodibility factor (*K*) of most of the savannah soils was low. For instance, *K* values for a tropical ferruginous soil from Sefa (Senegal) ranged from 0·04 to 0·17. Charreau also reported generalized runoff losses from different countries (Table 5.5 p. 153) and under different crops (Table 5.2).

Table 5.2. Amount of runoff under various crops at Sefa, Senegal (Charreau, 1972)

Vegetation	Rainfall (mm/an)	Runoff (%)	Soil loss (tonnes ha^{-1}/an)
Natural bush	1167	0·9	0·2
Groundnuts	1241	22·8	6·9
Sorghum	1113	34·1	8·4
Maize	1279	30·9	10·3

5.1.2. FOREST REGION

The problem of soil erosion in the forest regions, where natural fallow systems of cultivation are the way of life, becomes serious only if the forest vegetation is completely removed and arable cultivation maintained for several years, so that forest regeneration is prevented (Plate 17).

Considerable work has been done in the forest region of the Ivory Coast at Abidjan (Roose, 1967, 1973). Roose has also made use of Wischmeier's erosivity index (*R*) and prepared an iso-erodent map to cover most of West Africa (Chapter 5.4). Erosion research work at IITA near Ibadan in Nigeria was initiated in 1970 (Lal, 1976).

At Ibadan, soil erodibility factors (*K*) have been found to vary between 0·06 and 0·36 depending on the time after forest removal and the tillage techniques adopted (Lal, 1976a). For bare fallow plots on 15 per cent slope the annual soil loss reached 250 tonnes/ha. of which 50 per cent was lost in a single storm (Table 5.3). Mulching minimized runoff and soil loss even on the 15 per cent slope. A recent survey of the soil erosion problem in the East Central State of Nigeria has been completed by Niger Techno Ltd. (1975). Soil erodibility (*K*) for various soils in East Central State was reported to vary from 0·12 to 0·48. *K* values for the major soils were close to 0·33. These soils in the forest region of West Africa are therefore highly erodible.

Table 5.3. Effect of slope and soil management on runoff (mm) and soil loss (tonnes/ha) in 1973 (Lal, 1976a). Rainfall in the first season = 781 mm; rainfall in the second season = 416 mm

Treatment	Soil and water loss at different slopes (%) Runoff (mm) 1	5	10	15	Soil loss (tonnes/ha) 1	5	10	15
(a) First season 1973								
Bare fallow	315·7	347·3	311·0	316·5	7·5	80·4	152·9	155·3
Maize–maize (mulch)	0·0	6·9	20·3	16·8	0·0	0·0	0·1	0·0
Maize–maize	55·1	158·7	52·4	89·9	1·2	8·2	4·4	23·6
Maize–cowpeas (no-till)	11·4	11·8	20·3	21·0	0·0	0·2	0·1	0·1
Cowpeas–maize	19·9	81·2	51·4	46·1	0·6	5·6	3·2	7·6
(b) Second season 1973								
Bare fallow	191·7	195·8	193·1	185·4	3·7	75·8	79·7	73·9
Maize–maize (mulch)	0·0	4·0	9·0	7·5	0·0	0·0	0·1	0·0
Maize–maize	9·1	65·o	36·0	71·4	0·4	2·8	2·8	17·1
Maize–cowpeas (no till)	6·0	6·5	10·2	10·7	0·0	0·0	0·1	0·0
Cowpeas–maize	22·2	64·5	30·5	105·4	0·3	4·0	3·0	35·4

5.1.3. ECOLOGICAL DIFFERENCES BETWEEN THE FOREST AND SAVANNAH REGIONS

Soil erosion on equivalent slopes is liable to be more severe in the savannah than in the forest regions. Some of the factors causing greater erosion problems are listed in Table 5.4.

Table 5.4. Differences between the forest and savannah regions affecting erosion severity

	Forest	Savannah
Grazing	Not serious	Usually extensive grazing
Burning	Not serious	Usually annual
Vegetation cover	Adequate throughout the year	Not adequate when overgrazing or drought occurs
Soil compaction	Not serious under natural forest cover	May be seriously compacted by animal traffic
Length of dry season	Short	Long
Predominant land form slopes	Convex	Concave

The land form in the savannah zone is, however, often less conducive to erosion than the widespread convex slopes of the forest areas. These slopes, stable under forest cover, are liable to erode rapidly once exposed. After the dry long spells in the savannah region, vegetation cover is sparse. Uncontrolled grazing and the universal practice of burning result in complete vegetation removal and compaction of the surface horizon (Plate 18). When the dry season ends, rains of high intensity falling on compacted and exposed surface soils of low infiltrability result in serious runoff and soil erosion. Sometimes the water draining from ridges or animal footpaths results in severe incisions (Plate 19), which mark the beginning of the growth of large, uncontrollable and devastating gulleys (Plate 20). Bare hills, which were once covered by protective vegetation, shed water which also starts the formulation of gulleys (Plate 21). Development projects such as mining and road construction may further aggravate the problems.

In the forest region there is little erosion until man intervenes by removing forest for arable cultivation. Mechanical methods of deforestation lead to the exposure of structurally unstable soils to intense tropical rainstorms. Even conventional erosion control practices such as bench contours and graded waterways are not always effective. Ridges, though beneficial for gentle storms, can lead to serious erosion hazards under heavy storms, if a ridge breaks. In parts of the savannah region the problem is already unmanageable; it can get out of hand elsewhere if land development projects are not approached cautiously.

REFERENCES

Adu, S. V., 1972, Eroded savanna soils of the Navrongo–Bawku area, Northern Ghana, *Ghana J. Agric. Sci.*, **5**, 3–12.

Brammer, H., 1962, Soils, in J. B. Wills (Ed.), *Agriculture and Land Use in Ghana*, Oxford University Press, pp. 88–126.

Brook, T. R., 1955, Soil and water conservation, *Mon. Bull. Coffee Bd. Kenya*, **20**, 231–265.

Chalk, A. T., 1963, Soil conservation in Northern Nigeria and a suggested programme, USAID Rep. C-34.

Charreau, C., 1965, Dynamique de l'eau dans deux sols du Senegal, *Agron. Trop. Paris*, **16**, 504–581.

Charreau, 1968, Rain and erosion, *Afr. Soils*, **13**, 241–244.

Charreau, C., 1969, Influence des techniques culturales sur le developpement du ruissellement et de l'erosion en Casamance, *Agron. Trop. Paris*, **24**, 836–842.

Charreau, C., 1972, Problèmes poses par l'utilization agricole des sols tropicaux par des cultures annuelles, *Tropical Soil Research Symp.*, IITA, Ibadan, May 1972.

Charreau, C., and L. Seguy, 1969, Mesure de l'erosion et du ruissellement a Sefa en 1968, *Agron. Trop. Paris*, **24**, 1–55–1097.

Duckar, H. C., and S. T. Heyle, 1948, Some studies on cultivation practices, food crops and the maintenance of fertility at the cotton station, Nyasaland, *Trop. Agric.* (*Ceylon*), **104**, 155.

Eksteen, L. L., and M. Van des Spuy, 1941, Effect of the soil mulch in South Africa, *Trop. Agric.* (*Ceylon*), **96**, 231.

Elwell, H. A., and M. A. Stocking, 1973a, Rainfall parameters to predict surface runoff yields and soil losses from selected field-plot studies, *Rhod. J. Agric. Res.*, **11**, 123–130.

Elwell, H. A., and M. A. Stocking, 1973b, Rainfall parameters for soil loss estimation in a subtropical climate, *J. Agric. Engng Res.*, **18**, 169–177.

Elwell, H. A., and M. A. Stocking, 1976, Vegetal cover to estimate soil erosion hazard in Rhodesia, *Geoderma*, **15**, 61–70.

FAO, 1965, *Agricultural Development in Nigeria*, 1965–80, FAO, Rome.

Floyd, B., 1965, Soil erosion and deterioration in Eastern Nigeria, *Niger. Geogr. J.*, **8**, 33–44.

Fournier, F., 1960, Climat et erosion. La relation entre l'erosion du sol par l'eau et les precipitations atmospheriques, These Doct. es-lettres, PUF, Paris.

Fournier, F., 1962, Carte du danger d'erosion en Afrique au sud du Sahara, CEE-CCTA.

Fournier, F., 1967, Research on soil erosion and soil conservation in Africa, *Afr. Soils*, **12**, 53–96.

Gethin-Jones, G. H., 1936, Conservation of soil fertility in coffee estates, with special reference to anti-erosion measures, *East Afri. Agric. J.*, **1**, 456–462.

Grove, A. T., 1949, Farming systems and erosion on some soils in the South-eastern state of Nigeria, *Bull. Agric. du Congo Belge*, **40**, 3–4, 2150–2155.

Grove, A. T., 1951, Land use and soil conservation in parts of Onitsha and Owerri provinces, *Geological Survey of Nigeria*, Bull No. 21, Nigerian Govt., Lagos.

Harris, F. E., 1948, Soil conservation in Rhodesia, *Trop. Agric.* (*Ceylon*), **111** (6), 358.

Hudson, N. W., 1957a, The design of field experiments on soil erosion, *J. Agr Eng. Res.*, **2**, 56–67.

Hudson, N. W., 1957b, Erosion control Res. *Rhod. Agric. J.*, **54**, 297–323.

Hudson, N. W., 1959, Results of erosion research in Southern Rhodesia, Advisory Leaflet 13, Federal Department of Conservation and Extension, Salisbury.

Hudson, N. W., and D. C. Jackson, 1959, Results achieved in the measurement of erosion and runoff in Southern Rhodesia, *Third Inter-Afri. Soils Conference*, *Dalaba*, CCTA, pp. 573–583.

Hutchinson, J., M. L. Manning and H. G. Farbrother, 1958, On the characterization of tropical rainstorms in relation to runoff and percolation, *Quart. J. Roy. Met. Soc.*, **84**, 250–258.

Jones, R. G. B., 1975, Report on a soil conservation consultancy to study soil erosion problems on the Jos Plateau, Land Resources Report No. 6, Ministry of Overseas Development, UK.

Jones, M. J., and A. Wild, 1975, Soils of the West African Savanna, Commonwealth Bur. Soils. Techn. Comm. No. 55.

King, H. C., 1939, Erosion—A Mauritian measure for protecting water course, *Trop. Agric.*, **XCIII**, No. 5.

Kowal, J., 1970a, The hydrology of a small catchment basin at Samaru, Nigeria. III. Assessment of surface runoff under varied land management and vegetation cover, *Niger. Agric. J.*, **7**, 120–123.

Kowal, J., 1970b, The hydrology of a small catchment basin at Samaru, Nigeria. IV. Assessment of soil erosion under varied land management and vegetation cover, *Niger. Agric. J.*, **7**, 134–147.

Kowal, J., 1970c, Effect of an exceptional storm on soil conservation at Samaru, Nigeria, *Niger. Agric. J.*, **7**, 163–173.

Lal, R., 1976a, Soil erosion investigations on an alfisol in Southern Nigeria, IITA Monograph I.

Lal, R., 1976b, Is soil erosion a serious obstacle towards arable farming in Africa?, African Regional Water Conference, 20–24th Sept. 1976, Addis Ababa, Ethiopia.

Le Mare, P. H., 1954, Tie-ridging as a means of soil and water conservation and yield improvement, *Proc. 2nd Inter. Afri. Soils Conf.*, *Leopoldville*, p. 595.

Maher, C., 1950, Soil conservation in coffee, *Mon. Bull. Coffee Bd. Kenya*, **15**, 283.

Niger Techno Ltd., 1975, Prefeasibility study of soil erosion in the East Central State of Nigeria, Vols. I and II, Technical SPA International General Engineering, Rome.

Plate 17. Serious gully erosion can result in devastation of the landscape if forest cover is removed

[Facing page 148

Plate 18. Uncontrolled grazing results in a scanty vegetative cover

Plate 19. Water concentrated in animals' footpaths can initiate severe gullying

Plate 20. A large gully initiated from a footpath

Plate 21. Eroded bare hills shed water and damage adjacent areas

Plate 22. Gullies have destroyed vast areas of land in Eastern Nigeria

Plate 23. If ridges are not tied runoff on water-ridge channels can initiate erosion

Plate 24. Mulches protect the soil from raindrop impact. This photograph illustrates how plant debris has protected the underlying soil, while surrounding topsoil has been removed to a depth of 5 cms

Plate 25. Clean weeding may result in serious erosion on sloping lands. Here the protection afforded by a planted cassava stake can be seen, while the clean weeded inter row area has been severely eroded

Plate 26. Mixed cropping provides good ground cover and helps protect the soil from erosion

Obeng, H., 1972, Country Report—Ghana, Tropical Soil Research Seminar, IITA, Ibadan, May 1972.

Ofomata, G. E. K., 1964, Soil erosion in the Enugu region of Nigeria, *Afr. Soils*, **9**, 289–348.

Ofomata, G. E. K., 1965, Factors of soil erosion in the Enugu area of Nigeria, *Niger. Geogr. J.*, **8**, 45–59.

Rapp, A., L. Berry and P. Temple, 1972, Studies of soil erosion and sedimentation in Tanzania, *Geografiska Annaler*, **54A**, Nos. 3 and 4.

Roche, L., 1973, The practice of agri-silviculture in the tropics with special reference to Nigeria, *FAO Soils Bull*, **24**, 170–190.

Rosenburg, H. J. Van, 1955, Runoff and soil erosion tests, Mpwapwa, Central Tanganyika, *E. A. Agric. J.*, **20**, 228–231.

Roose, E., 1967, Dix annees de mesure de l'erosion et du ruissellement au Senegal, *Agron. Trop., Paris*, **22**, 123–152.

Roose, E., 1973, Dix-sept annees de mesures experimentales de l'erosion et du ruissellement sur un sol ferralitique sableux de basse Cote d'Ivoire, Thesis pour Docteur-Ingenieur, Univ. d'Abidjan, Pub. ORSTOM, Abidjan.

Roose, E., and R. Bertrand, 1971, Contribution a l'etude de la methode des bandes d'arret pour lutter contre l'erosion hydrique en Afrique de l'Ouest. Resultats experimentaux et observations sur le terrain, *Agron. Trop., Paris*, **26**, 1270–1283.

Saunders, A. R., 1944, Cultivation and soil mulch in South Africa, *Trop. Agric.* (*Ceylon*), **100**, 118.

Sperow, Jr, C. B., and R. F. Keefer, 1975, An introduction to soil science applied to East Africa, West Virginia Univ. Publication.

Stockdale, F., 1937, Soil erosion in the colonial empire, *Empire J. Exp. Agric.*, **5** (20). (20).

Thompson, A. W., 1939, Some aspects of soil conservation, *East Afr. Agr. J.*, **4**, No. 4. 4.

Verney, R., and P. Williams, 1965, Results of studies of erosion on experimental plots in Dahomey, Proc. OAU/STRC Symp. on the Maintenance and Improvement of Soil Fertility, Khartoum.

Wallis, J. A. N., 1961, The place of herbicides in the management of Kenya Coffee, *Kenya Coffee*, March 1961, 77.

Wrigley, G., 1969, *Tropical Agriculture*, Faber & Faber, London.

5.2

Farming Systems and Soil Erosion in West Africa

B. N. OKIGBO

5.2.1. EROSION HAZARD IN WEST AFRICA

Adverse effects of erosion on soils, vegetation, ecology and human welfare have received the attention of various authors (Grove, 1951, 1952; Guillaúme, 1951; Jurion, 1956; Langdale-Brown, 1968; Middleton, 1952; Strahler, 1956; FAO, 1960, 1965; Fournier, 1967; Floyd, 1969; Ahn, 1970), and space and relevance of this to the topic under consideration does not permit a detailed treatment here. Insofar as erosion causes general soil degradation and loss of soil fertility and productivity, it constitutes a hazard to farming. Equally, farming is one of the major human activities which when not properly conducted constitutes a hazard to the soil. Thus, according to FAO (1960, 1965), farming is most hazardous in relation to the extent of damage and potential damage due to erosion in tropical, mountainous and hilly regions of the world where heavy rains are common. In these areas, it is difficult to control water erosion on sloping land. The areas in Africa where serious damage has been observed include Swaziland, Lesotho, Botswana, East Africa, heavily populated areas of Ruwanda and Burundi, southeastern and Bauchi plateau areas of Nigeria, parts of northern Ghana and hilly areas of Sierra Leone.

In general, in West Africa as elsewhere in the tropics, water erosion, especially of the sheet erosion type, is more widespread in unprotected croplands and overgrazed pastures than wind erosion. This kind of erosion is more serious than is usually appreciated because it slowly and gradually depletes the soil qualitatively and quantitatively and sets the stage for more spectacular gully erosion and floods (FAO, 1965). Lal (1973) related erosion hazard to rainfall distribution in West Africa and outlined the parts of West Africa that are particularly susceptible to rainfall erosion. The areas concerned included all of Zaire, Gabon, the Cameroun Republic, Nigeria, Ghana, Dahomey, Ivory Coast, Liberia and Sierra Leone, with a northern zone consisting of areas with average annual rainfall of less than 1000 mm and a southern zone consisting of areas with more than 1000 mm annual rainfall, that are susceptible to rainfall erosion when the vegetation cover is removed.

Observations under experimental conditions in various parts of West Africa also support the view that sheet and to some extent rill erosion are serious in unprotected soil (Mouttapa, 1973; Lal, 1973). Estimates of quantities of soil removed by erosion from soils under different slopes and surface cover in Ouagadougou (Upper Volta), Sefa (Senegal), Bouake and Abidjan (Ivory Coast) are presented in Table 5.5.

A more impressive manifestation of the seriousness of the erosion problem in parts of West Africa is the spectacular and disastrous gullies which have destroyed vast areas of land in the Nanka, Agulu, Oko and Enugu areas of the Anambra State and Shendam and western Pankshin areas of the Plateau State in Nigeria (Grove, 1951, 1952; Floyd, 1969; Udo, 1970; Ofomata, 1964) (Plate 22). Recently (Anon, 1974a) the erosion damage in East Central State was estimated as causing losses of over 13 million tonnes of soil, or amounting to over ₦300 million annually, with the gullies extending to depths of over 120 m and up to 2 km wide in places.

5.2.2 EROSION HAZARD AND FARMING SYSTEMS IN WEST AFRICA

The current concern about drought, erosion and soil conservation in various parts of Africa is only an aspect of a long-standing concern. The gullied hillsides and badlands of the Anambra State of Nigeria, already referred to above, were noted by European visitors as early as 1913 (Grove, 1951). An enquiry concerning the possible drying up of land in the then French West Africa was reported by Hubert in *Bulletin de l'IFAN* in 1917 (Anon, 1957). The Inter-African Information Bureau for Soil Conservation and Land Utilization which started the journal *African Soils* was set up in 1948 with the objectives of (i) providing a medium for exchange of information among those interested in research on soils, erosion control and soil conservation, in addition to (ii) promoting the adoption of sound agricultural methods of soil conservation. In fact, it would appear that, as indicated in reviews of research and related activities on erosion, soil conservation and land utilization in the then British Colonial Africa (Greene, 1951), the Intertropical zones of the French Union (Guillaume, 1951), the Belgian Congo (Jurion, 1951), the Union of South Africa (Adler, 1951) and Africa and Madagascar (Fournier, 1967), there was perhaps greater concern about these problems then than there is today. The early reviews and reports of several missions sent out to parts of Africa, on soil conservation and erosion control, emphasized the importance of human activities, especially those associated with past and present farming systems, in the evolution of present landscapes and soil degradation due to erosion (Guillaume, 1951; Porteres, 1952a, b; Guilloteau, 1955).

Of the various factors which determine the intensity and seriousness of erosion, vegetation cover, land use and farming are to a large extent dependent on human activities. The human activities which affect erosion are listed in Table 5.6. Each of the farming systems in Table 5.7 usually involves one or more of these activities. The ways in which they influence erosion are considered below.

Table 5.5. Runoff and erosion under various covers of vegetation in parts of West Africa. Source: Charreau (1972) in Mouttapa (1973)

Locality	Study period	% slope	Average annual rainfall (mm)	Annual runoff, % Forest	Crop	Barren soil	Erosion (tonnes/ha) Forest	Crop	Barren soil
Ouagadougou (Upper Volta)	1967–70	0·5	850	2·5	2–32	40–60	0·1	0·6–0·8	10–20
Sefa (Senegal)	1954–68	1·2	1300	1·0	21·2	39·5	0·2	7·3	21·3
Bouake (Ivory Coast)	1960–70	4·0	1200	0·3	0·1–26	15–30	0·1	1–26	18–30
Abidjan (Ivory Coast)	1965–70	7·0	2100	0·1	0·5–20	38	0·03	0·1–90	108–170

Table 5.6. Farm practices and operations determining the extent of erosion hazard

1. Bush-clearing (slashing, pruning, burning, mechanical and chemical methods)
2. Burning of vegetation for ceremonial purposes, flushing out small game, encouraging green bite for livestock, etc.
3. Preplanting cultivations (flat, ridge, mounds, ploughing)
4. Fallowing and mulching
5. Manuring and fertilization
6. Weeding
7. Cropping systems (row, sole, intercropping and pasturing)
8. Logging and related forestry activities
9. Harvesting and grazing
10. Restriction of livestock
11. Farm and household refuse disposal
12. Footpaths to farms, water supplies and springs, and markets
13. Compound development and maintenance

Table 5.7. General classification of farming systems found in West Africa. Adapted with modifications from Greenland (1973), Floyd (1969), Morgan and Pugh (1969), and Laut (1971)

A. Traditional and transitional farming systems
   1. Shifting cultivation (phase I including nomadic herding), $L > 10$
   2. Bush-fallowing or land rotation (shifting cultivation phase II), L 5–7
   3. Rudimentary sedentary agriculture (shifting cultivation phase III), L 3–5
   4. Compound farming and intensive subsistence agriculture (shifting cultivation phase IV), $L < 2$
   5. Terrace farming and floodland agriculture, $L \leqslant 1$

B. Modern farming systems and their local adaptations
   1. Mixed farming
   2. Livestock ranching
   3. Intensive livestock production (poultry, pig production and dairying)
   4. Large-scale plantations
      (a) large-scale food and arable crop farms
      (b) Irrigation projects
      (c) Large-scale tree crop plantations
   5. Specialized horticulture
      (a) Fruit trees and/or ornamentals
      (b) Market gardening

L = land use factor of Allan (Greenland, 1973) = $(C+F)/C$ where C = cropping period in years and F = fallow period in years.

1. *Bush-clearing*. This is the most important and first operation carried out to facilitate cultivations and planting. In traditional farming systems hand-clearing methods are generally used. Most of the shrubs are cut at heights of about a metre from the ground, but where the stumps are to be used for staking they are cut at heights of 2–3 m. Usually very large trees are left standing, but where they are very close together the branches are sometimes heavily pruned down. The clearing is usually carried out in the dry season, and completed by burning the dried materials. The stumps are usually not dug out except when clearing to establish plantations, when the stumps close to the planting holes are usually dug out with picks. After two to three years when cropping is completed, the stumps sprout and regenerate and grow into secondary bush. In such situations no serious erosion occurs except during the short period between cultivations, planting and development of crops, which should provide some cover during the growing season.

Clearing for plantations may be carried out by hand where labour is cheap, but usually it may be accomplished by chemical and mechanical methods. In chemical clearing, ring-barking is done at convenient heights from the ground and chemicals such as sodium arsenite or 2,4,5-T are used to poison the plant (Webster and Wilson, 1966). Winches may also be used in clearing, but usually for large areas of thick forest almost all operations can usually be mechanized. Both the clearing of the undergrowth and the felling of trees may be done with a bulldozer. Heavy tractors could be employed in chain-clearing. For tree crop plantations, stumping is not necessary and all the plant materials are left to dry before burning. Some trees are selectively left to provide shade, but where annual crops are to be grown, mechanical stumping, piling in windrows and root cutting may be carried out to facilitate mechanical cultivation. Unless carefully done, bulldozer piling, stumping and root cutting may result in serious disturbance of the soil surface, exposure of subsoil in places and destruction of soil structure by compaction. This kind of clearing often results in serious erosion leading to crop failure except where it is carefully carried out without disturbance of the surface soil and when the soil is not very wet.

2. *Burning of vegetation.* The cut and burn technique is the traditional method of bush-clearing in West Africa. Burning is also carried out for various reasons including flushing out of small game, ceremonial purposes or for fun, encouragement of 'green bite' or fresh uniform growth of fodder for livestock, and sometimes by accident (Grove 1951, 1952; Floyd, 1969). Burning may be advantageous in releasing nutrients to crops and other plants, and in the destruction of weed seeds, pathogens and pests. However, it causes nutrient losses through leaching and surface runoff, destruction of soil organic matter and structure, and destruction of fauna and flora, thereby disrupting the ecological balance which existed previously to an extent that makes restoration impossible. Sometimes valuable economic trees are destroyed by fire or exposed to secondary infection by pathogens. Burning *per se* is not altogether harmful, and controlled early burning of grasslands in the savannah areas is known to encourage tree growth and prevent late destructive burning (Vine, 1968). Frequent burning of forest and woodlands and constant clearing of bushes and felling of trees for fuel constitute practices which expose the soil to erosion.

3. *Preplanting cultivations.* In traditional farming systems, use of preplanting cultivations such as mounds, heaps, ridges and beds is widespread all over West Africa, especially on shallow soils and soils with high water-table during certain months of the year. Only on flat well-drained soils, alluvial plains and deep sandy loams is planting on the flat common. Heaps, mounds and ridges constructed on bare soil and sometimes running up and down the slope may expose the soil to serious erosion by concentrating the runoff so that it develops rills and gullies (Ahn, 1970; Floyd, 1969) (Plate 23). Tying of ridges (Plate 7), mounds and heaps, orienting them parallel to the contour, and mulching may reduce the danger of erosion (Plate 24).

4. *Fallowing.* Fallowing involving natural bush regrowth or planted fallows with selected plants is used in traditional agriculture to rest the soil, recycle nutrients and thereby restore the soil fertility for another cycle of cropping for one or more years. Fallows are most effective when they are long enough (7–20 years) to bring sufficient nutrients to the surface through leaf fall. When fallows are short, soil fertility is not adequately restored and subsequent crops and fallows fail to grow well and provide adequate cover for the soil. In modern agriculture, fallows of shorter duration may be planted with cover crops or green manure crops which serve to restore soil structure and add some organic matter to the soil.

5. *Manuring and fertilization.* Manuring is used in market gardening and to some extent in intensive farming in house compounds to maintain soil fertility

and productivity. Fertilizers are used for the same purpose in modern agriculture. Fertilizer use is not widespread in traditional agriculture and the little that is carried out is mostly restricted to cash or export crops rather than food crops. This is because fertilizers are expensive, frequently fail to give significant returns on land that has been in fallow for several years and, where they do give responses, the food crops fail to command a sufficiently attractive price to make their use economic.

6. *Weeding.* Clean weeding exposes the soil to erosion unless the crop canopy is thick and close enough to protect the soil adequately. Weed suppression by mulching, by close cropping, and by suitably selected combinations of inter- and relay-cropping is much more desirable than other systems of weeding. Weeding by cultivation, even in plantation crops, especially where the soil is more erodible and on sloping land, may cause serious erosion (Plate 25).

7. *Cropping system.* In traditional agriculture, intercropping of arable crops and tree crops, some of which are planted and harvested at different times, ensures that the soil is usually protected (Plate 26) by a vegetative cover. Row crops such as maize, planted at wide spacing, expose the soil more to sheet erosion than non-row crops such as cowpeas, forage grasses and sweet potatoes, which are planted at closer spacing and provide adequate soil cover. While mono-cropping with maize can lead to severe erosion, this is much reduced if it is intercropped with cowpeas, cassava or sweet potatoes.

8. *Logging and forestry activities.* Tree-felling and logging with heavy equipment may result in compaction of the soil, especially along bush paths and roads. These form channels for gully erosion, which is most serious when they run straight up and down the slope.

9. *Grazing.* Natural and sown pastures can provide adequate cover to the soil, but overgrazing and overstocking of pastures may result in removal of cover. Added to this is the trampling by livestock, which converts their trails and footpaths to channels for gully erosion. Rotational grazing, controlled burning and good range management eliminate this potential erosion hazard.

10. *Restriction of livestock.* In traditional agriculture, livestock are often allowed to wander without restriction, so that they cause damage to crops and vegetation. Goats are notorious in this regard, and their activities may constitute a serious erosion hazard even in the humid tropics, especially if their numbers and movements are not controlled.

11. *Farm and household refuse.* These in many rural areas of West Africa constitute the manure and mulching material with which soil fertility is maintained. Their use as manure or mulch ensures continuous cultivation and ground cover in compound farms and gardens.

12. *Footpaths to farms, springs and markets.* Grove (1952) in his study of erosion and soil conservation in the Oko area of the Anambra State, Nigeria, indicated that footpaths, some of which run up and down the slope, and especially those leading to springs at the foot of the slope, farms, markets and compounds, amounted to about 5 per cent of the total area. These are always kept clean and bare, and so form the channels from which gully erosion develops.

In considering the association of farming systems with erosion hazard, each farming system may be considered as an agricultural enterprise satisfying specific objectives and involving various kinds of plant and/or animal inputs, together with the practices and operations with which it is managed in a given environmental setting. The various agricultural enterprises (farming systems) currently found in West Africa are listed in Table 5.7. They vary in their objectives, complexity, degree of sophistication and in the interaction of these practices with climatic and other environmental factors. The common practices often encountered in the farm business and in the farmer's environment are listed in Table 5.6. The extent to which these may affect erosion has been reviewed above. Normally, in any one situation man evolves his practices through centuries of experience so that they interact with the environment in such a way as to maintain a stable system of production. Any change in the social and economic factors that affect the farming system and its objectives is usually also associated with changes in practices and in the plant and animal inputs used in a given set of circumstances.

Under traditional natural fallow systems, where fallow periods are still long enough (up to 10 years or more) a reasonable balance can be maintained between fertility loss during cultivation and restoration under fallow. Erosion is minimized by ensuring that there is some cover during clearing, and by growing several crops together which protect the soil and minimize weeding. According to Grove (1952), before Western influence reached West Africa, ecological balance was maintained between people and their environment by controls of inter-tribal wars, famine, diseases, and farm practices and customs based on centuries of experience, adapted to maintain the relatively low populations.

In the last hundred years, various aspects of modernization, especially in terms of economic and demographic expansion, have brought about changes which have increased population pressure on the land and disrupted the balance of the traditional farming systems by seriously shortening the periods of fallow. Associated with this is the adoption of ill-advised and irrational agricultural practices which, while appropriate elsewhere, are incompatible with the environmental conditions and resource endowments of the traditional farmer of the humid tropics. The changes which have resulted in the present predicament include:

(i) Increasing population density and pressure on agricultural land and consequent increase in area of cultivated land, sometimes spreading to marginal areas that are unsuitable for farming with traditional techniques (Gourou, 1956).

(ii) More and more effort being devoted to increase food and export crop production to satisfy increased demand (Guillaume, 1951).

(iii) Overgrazing resulting from increased livestock populations and stocking rates (Guillaume, 1951; Grove, 1951).

(iv) Widespread deforestation resulting from shortening cycles of cropping and fallowing.

(v) Migration of peoples to areas where their traditional methods of agricultural production are unsuitable to the existing environmental conditions (Porteres, 1955; Guilloteau, 1955; Gourou, 1956).

(vi) Rate of development of agricultural techniques and organization of space not keeping pace with the increased rate of change from extensive farming techniques to more intensive systems (Gourou, 1956).

(vii) General economic development and improvement of means of communication which have brought West Africa closer to markets and drastically increased the pace of commercialization of agriculture (Gourou, 1956; Guillaume, 1951).

(viii) Increasing use of mechanical cultivation and large-scale farming (Obeng *et al.*, 1973).

Various classifications of agricultural or farming systems have been reported (Ramsey, 1968; Floyd, 1969; Morgan and Pugh, 1969; Laut, 1971; Greenland, 1973). The list in Table 5.7 is an adaptation from the various systems studied to suit the West African situation. It is by no means an exhaustive list. A brief summary of the characteristics of these farming systems and the extent of erosion hazard associated with them is presented below.

1. *Shifting cultivation* (*phase I including nomadic herding*)

(i) With the exception of nomadic herding, this system is not widespread in West Africa, but confined to a few isolated, semi-arid areas.

(ii) Non-sedentary, use of simple houses or movable tents, migration follows seasonal advance and retreat of rainfall.

(iii) An extensive system of agricultural production involving mainly livestock with sometimes (but very rarely) some cereals in favoured situations.

(iv) With improved veterinary services, increasing livestock numbers result in overstocking which is associated with increased erosion hazard, especially around drinking holes and watering places.

2. *Shifting cultivation* (*phase II*) *or bush-fallowing*

(i) Farms small in size often in isolated patches in secondary woodland or, rarely, virgin forest.

(ii) Homes usually permanent with small areas of more or less continuously cropped compound gardens around them, outlying farms change position frequently.

(iii) Cut and burn using simple tools practised at start of cultivation cycle.

(iv) A range of crops grown in mixed cropping systems which may include tree crops in the compound gardens.

(v) Some small numbers of livestock may be involved.

(iv) Length of bush fallows after each 2–3 years cropping period reasonably long (>7 years) but may vary with population pressure.

(vii) Erosion hazard low except along footpaths, and in compounds.

3. *Rudimentary sedentary agriculture* (*shifting cultivation phase III*)

(i) Farm size smaller than in 2 above and less than 50 per cent of community land in fallows which rotate after each cropping cycle.

(ii) Cut and burn technique using simple tools followed.

(iii) Permanent settlement with compound gardens more intensively cultivated and associated with systems of outlying fields at varying distances from homestead.

(iv) Range of crops grown in mixed cropping systems and some livestock associated with them.

(v) Some small plots of protected or improved perennial cash crops often involved.

(vi) Planted, protected or natural bush fallows may be used in outlying fields with lengths of fallow 7–8 years or less.

(vii) Erosion hazard slightly increased especially with decreases in frequency and duration of fallow.

4. *Intensive sedentary agriculture or compound farming (shifting cultivation phase IV)*

(i) Farm size small, holdings fragmented.

(ii) Permanent settlements, simple tools and limited cut and burn techniques.

(iii) Compound farms more intensively and continuously cropped with fertility maintained with applications of household, farm and animal waste.

(iv) Crop mixtures including some perennial cash crops and fruit trees—a vegecultural complex with some livestock.

(v) Length of fallow in outlying farms often less than four years.

(vi) Erosion hazard high on slopes, footpaths and compounds.

5. *Terrace farming and floodplain agriculture*

Remnants of these according to Gourou (1956) and Morgan and Pugh (1969) are found in the Atacora Mountains, Bauchi Plateau, Mandara Mountains and Adamawa Highlands. They are located where people took up defensive positions on hillsides and developed very intensive systems of terrace farming. Some have since been abandoned, with the people reverting to shifting cultivation in the lowlands. Elaborate terrace farming systems are still being used and maintained at Maku, in the Awgu area of Anambra State in Nigeria. Upper slopes (A) and floodplains (B) are both used in this way.

A

(i) Farm sizes are small, simple tools used but settlement permanent.

(ii) Terraces built of stones roughly along the contours with adequate or good water management techniques.

(iii) Fertility of soil maintained with farm and household refuse and animal waste.

(iv) A range of crops grown in mixed or monoculture and rotated—very intensive system with vegetables of major importance.

(v) Very good use made of different topographic situations or soils at different elevations with tree crops on top and some in valley bottoms, annual staples on terraces in between.

(iv) Population pressure very high and migration rate also high.

(vii) Erosion hazard high on footpaths along slopes and where terrace walls break down.

B

(viii) Floodplain agriculture takes place on very rich alluvial soils and sometimes on hydromorphic soils.

(ix) Very high yields of yams, bananas and rice are produced. Often the farmer has to race with the flood and crops may be lost unless there is good water control.

(x) Fishing may be a very important occupation.

(xi) On hydromorphic soils crops and vegetables are produced in the dry season.

(xii) Erosion hazard not high on the flat lowlands, except that associated with floods, especially on slopes.

6. *Mixed farming*

(i) Mainly involves production of a range of livestock (cattle, sheep, goats, etc.) in combination with systems of cultivation of crops.

(ii) Consists mainly of two kinds: (a) traditional types found among the Serere of Senegal, where it is declining with increased interest in groundnuts, the Fulani of the Foutta Djallon area of Guinea and Shuwa Arabs of northeastern Nigeria, and (b) more modern aspects of mixed farming introduced into areas such as Kano (Nigeria) in the 1940s with livestock kept for work and manure.

(ii) Soil fertility usually maintained around homesteads and on farms with farmyard manure.

(iv) Erosion hazard very high, especially in the savannah where there is no adequate cover at the end of the dry season.

7. *Livestock ranching*

(i) Large-scale ranching projects such as the one at Obudu in southeastern Nigeria have or are currently being developed in different parts of West Africa.

(ii) Erosion hazard usually associated with poor range management, overstocking and overgrazing. Gully erosion most serious along animal trails and waterholes. With increasing experience and stable economic conditions, management should improve, and erosion hazard be reduced.

8. *Intensive livestock production*

(i) These projects involve the rearing of cattle, pigs, poultry and dairy animals in intensive systems.

(ii) Erosion hazard is absent or limited where animals are fed indoors.

(iii) Where they are grazed, trampling of vegetation and grazing may expose some areas to erosion.

9. *Large-scale plantations*

(i) Many kinds of large-scale plantations owned by private individuals, cooperatives and governments are being developed in many parts of West Africa.

(ii) Of the three listed in Table 5.7, the lowest erosion hazard is associated with tree crop plantations such as rubber and oil palms, especially where they are grown with cover crops such as *Pueraria phaseoloides*.

(iii) The highest erosion hazard is associated with large-scale mechanized food and arable crop farms involving row crops (Obeng *et al.*, 1973).

(iv) Baldwin (1957) gives a very interesting example of this in *The Niger Agricultural Project*, at Mokwa, the failure of which he regarded as 'the most disastrous consequence of lack of adequate preliminary knowledge'. Very spectacular erosion gullies 1·8 m wide, 0·9 m deep and 1·6 km long resulted after two years, from inadequate attention to the erosion hazard. The same mistakes are repeated in many of the schemes now being introduced.

10. *Specialized horticulture*

(i) Market gardens and large-scale horticultural projects are developing in different parts of West Africa, especially close to urban centres.

(ii) As an aspect of intensive production systems, mulching of beds, construction of terraces and tied beds, etc., have often been done as regular practice.

(iii) Where beds are not protected by mulching and not aligned parallel to the contour, erosion hazard may be serious.

(iv) Moreover, certain types of irrigation systems on sloping land may introduce erosion problems.

(v) Where fruit trees are grown in plantations with cover crops, erosion is very much minimized.

5.2.3. CONCLUSIONS

On the basis of the above review of the erosion hazards in different farming systems in West Africa and their associated farming practices or operations, it is concluded that:

(i) traditional systems of soil management do not involve serious erosion hazard as long as the periods of fallow are reasonably long, vegetation cover is maintained and population pressure is low;

(ii) with increasing population pressure, overgrazing or change to more intensive systems of production on light soils and slopes with methods developed for extensive agricultural systems, erosion becomes a serious problem;

(iii) mechanization of clearing and cultivations still presents problems, especially on large-scale farms for growing row crops.

(iv) changes in traditional farming systems require appropriate changes in crop and/or animal inputs based on sound principles of soil conservation; and

(v) it is not only in farming but also in various aspects of multiple land use that sound conservation practices and protection of the environment should be emphasized.

Erosion has been a world-wide problem for centuries and still remains so in West African agriculture. It can be minimized by making everyone aware of the problem, by the enforcement of sound soil conservation legislation where necessary and feasible, and by the development of scientific water, soil and crop management techniques for catchments or watersheds of different sizes and for various topographic situations. The lowlands and hydromorphic soils of the humid tropics of West Africa hold high potentials for increased agricultural production without soil erosion and degradation, but the development of the appropriate technology for achieving this has received too little attention.

More effort and resources should be devoted to the prevention of erosion through research aimed at the development of better land clearing, water and soil management techniques for continuous cropping systems. This can only be achieved through the cooperation of scientists in different disciplines and across international boundaries.

REFERENCES

Adler, E. D., 1951, Soil conservation in the Union of South Africa, *African Soils*, **I** (1), 25–35.

Ahn, P. M., 1970, *West African Soils*, West African Agriculture Vol. 1, Oxford University Press, London.

Anon, 1957, Enquiry concerning the drying up of land in French West Africa, Climatological Abstracts, *African Soils*, **1** (1), 91–94.

Anon, 1974a, Erosion causes havoc, *Daily Times* (Lagos), January 24, 1974, p. 1.

Anon, 1974b, Problem of soil erosion, *Daily Times* (Lagos), Editorial, January 28, 1974.

Baldwin, K. D. S., 1957, *The Niger Agricultural Project* Blackwells, Oxford.

Baver, L. D., 1940, *Soil Physics*, Wiley, New York.

FAO, 1960, *Soil Erosion by Wind and Measures for its Control on Cultivated Lands*, FAO Agricultural Development Paper No. 71, FAO, Rome.

FAO, 1965, *Soil Erosion by Water, Some Measures for its Control on Cultivated Lands*, FAO Agricultural Development Paper No. 81, FAO, Rome.

Floyd, B., 1969, *Eastern Nigeria*, Macmillan, London.

Fournier, F., 1967, Research on soil erosion and soil conservation in Africa, *African Soils*, **22** (1), 53–96.

Gourou, P., 1956, The quality of land use of tropical cultivations, In W. L. Thomas, *Man's Role in Changing the Face of the Earth*, University of Chicago Press, Chicago, pp. 336–349.

Greene, H., 1951, Soil conservation in British Colonial Africa, *African Soils*, **1**, 13–19.

Greenland, D. J., 1973, Evolution and development of shifting cultivation, *Shifting Cultivation and Soil Conservation in Africa*, FAO Soils Bulletin, No. 24, FAO, Rome, pp. 5–13.

Grove, A. T., 1951, Land use and soil conservation in parts of Onitsha and Owerri Provinces, *Geological Bulletin of Nigeria*, No. 21, Zaria.

Grove, A. T., 1952, Land use and soil conservation in the Jos Plateau, *Geological Survey of Nigeria Bulletin*, No. 22, Zaria.

Guillaume, M., 1951, Soil conservation in the inter-tropical soil of French Union, *African Soils*, **1** (1), 21–24.

Guilloteau, J., 1955, The improvement of agricultural economy in Senegal, summary of Professor Porteres' Mission Report, *African Soils*, **3**, 11–44.

Jurion, P., 1956, Soil conservation in the Belgian Congo, *African Soils*, **4,** 35–44.

Kittredge, J., 1973, *Forest Influences*, Dover, New York.

Lal, R., 1973, Soil erosion and shifting agriculture, *Shifting Cultivation and Soil Conservation in Africa*, FAO Soils Bulletin, No. 24, FAO, Rome, pp. 48–71.

Langdale-Brown, I., 1968, The relationship between soils and vegetation, in R. P. Moss, *The Soil Resources of Tropical Africa*, Cambridge University Press, London.

Laut, P., 1971, *Agricultural Geography*, Vol. 1, 1st edn, Nelson, London.

Middleton, H. E., 1952, Modifying the soil physical system: modification of the physical properties of virgin soil in preparing it for cultivation, In B. T. Shaw, *Soil Physical Condition and Plant Growth*, Vol. 11, Agronomy Monographs, Academic Press, New York, pp. 24–41.

Morgan, W. B., and J. C. Pugh, 1969, *West Africa*, Methuen, London.

Mouttapa, F., 1973, Soil aspects in the practice of shifting cultivation in Africa and the need for a common approach to soil and land resources evaluation, *Shifting Cultivation and Soil Conservation in Africa*, FAO Soils Bulletin No. 24, FAO, Rome, pp. 37–47.

Ofomata, G. E. K., 1964, Soil erosion in the Enugu region of Nigeria, *African Soils*, **9**, 321–348.

Portères, R., 1952, Linear cultural sequences in primitive systems of agriculture in Africa and their significance, *African Soils*, **2**, 15–28.

Portères, R., 1955, Agricultural crisis in Senegal, *African Soils*, **3**, 41–52.

Ramsey, D. M., 1968, Land use in Nigeria, *Samaru Agricultural Newsletter*, **10** (4), 73–81.

Russel, E. W., 1968, Some agricultural problems of semi-arid areas, In R. P. Moss, *The Soil Resources of Tropical Africa*, Cambridge University Press, London.

Shaw, B. T., 1952, (ed.), *Soil Physical Condition and Plant Growth*, Academic Press, New York.

Strahler, A. N., 1956, The nature of induced erosion and aggradation, in W. L. Thomas, *Man's Role in Changing the Face of the Earth*, University of Chicago Press, Chicago.

Udo, R. K., 1970, *Geographical Regions of Nigeria*, Heinemann, London.

Vine, H., 1968, Developments in the study of soils and shifting agriculture in tropical Africa, In R. P. Moss, *The Soil Resources of Tropical Africa*, Cambridge University Press, London, pp. 89–119.

Webster, C. C., and P. N. Wilson, 1966, *Agriculture in the Tropics*, Longmans, Green, London.

5.3

Erosion Hazard and Farming Systems in East Africa

PETER M. AHN

'Soil erosion is nearly always caused by an unsuitable method of agriculture being practised, and since among primitive peoples the system of agriculture affects their whole way of life, methods of control involve . . . altering the whole social outlook.' E. W. Russell, *Soil Conditions and Plant Growth*, 10th edn., p. 772.

5.3.1. FUNDAMENTAL FACTORS AFFECTING EROSION HAZARD

Assessing erosion hazard is generally accepted as being a complex problem, but in East Africa it is probably much more complex than usual due to the very great variety of altitude and climate, of soils, and of farming systems found. As elsewhere, however, it is helpful to distinguish the effects of the following sets of factors.

(a) Climate, and particularly its erosivity.
(b) Soil, particularly its erodibility, which is related to inherent soil properties, including structural strength and rainfall acceptance, and degree and length of slope on which the soil rests.
(c) The cover provided by the vegetation or the crop, and the soil management.

Of these three factors, the last, the farming system and degree of cover provided at different times of the year, is probably the most important. It is also the most varied and difficult to examine quantitatively.

5.3.2. CLIMATE AND EROSION HAZARD IN EAST AFRICA

Although wind erosion is probably important in some of the drier parts of East Africa, little attention has been given to it. Stephens (1971) considers it significant in some areas of Uganda, particularly in the bare, dry plateau of eastern Karamoja, in denuded areas of Teso, and in the dry southwest of Toro, where fluffy volcanic soils are cultivated mechanically. Nevertheless, in East Africa in general water is undoubtedly the main agent of soil erosion.

In East Africa, experience has suggested that although tropical rain is more erosive in general than temperate rain, little or no erosion occurs with rainfall at low intensity (Chapter 3.1).

Climatic studies in East Africa of recent years have mainly emphasized rainfall reliability and confidence limits (Lawes, 1969; Brown and Cocheme, 1969; Braun, in preparation), but Lumb (1971) has calculated probable maximum precipitation for periods up to 24 hr, and Taylor and Lawes (1971) give a map of the actual 24-hr maxima recorded so far. Although rainfall amount generally increases with altitude, the storms of the lower altitude and generally lower rainfall areas are often more intense. This is indicated by the following calculated estimated maxima for a 10-km area (Table 5.8) (Lumb, 1971).

Table 5.8.

Altitude	1 hr	3 hr	6 hr	12 hr	24 hr
900 m	130 mm	260 mm	300 mm	350 mm	400 mm
1500 m	110 mm	220 mm	250 mm	290 mm	330 mm

Once in two years, maxima for a 24-hr period are, in ascending order of altitude, Mombasa 190 mm, Voi 152 mm, Kisumu 132 mm, Kabete 118 mm and Kitale 95 mm (Taylor and Lawes, 1971). Thus the lower areas of East Africa with generally lower rainfalls nevertheless experience the most intense storms, and this, together with the lack of natural cover in these areas, accounts for the often much greater erosion hazard. However, it appears that so far no detailed studied on continuous recording rainfall records have been carried out in order to calculate erosivity indices for different areas, and thus to provide an accepted general measure of erosivity.

5.3.3. THE ERODIBILITY OF EAST AFRICAN SOILS

The relation between aggregation and erodibility is discussed in Chapter 2.2. Many tropical African soils, particularly those developed from basic rocks, have a very stable microaggregation which makes them feel like loams or even sandy loams even though they may contain 60–80 per cent clay (Ahn, 1968). The widespread deep red permeable soils developed over trachytes and similar lavas in the Kenya highlands form an excellent example—they were in fact long referred to as Kikuyu loams despite the fact that they often have 70–80 per cent of clay (they are now better called the Kikuyu friable clays). Analyses by the author of a deep coffee soil of this type sampled primarily for work on forms of phosphate (Keter, 1974) indicate that the 'natural' texture obtained after destruction of organic matter with H_2O_2 and end-over-end shaking for four hours without dispersing agent was a silt loam or loam, whereas the true texture was a clay throughout, with 51–77 per cent clay. The following results (Ahn, Table 5.9) indicate that in the natural soil most of the clay was aggregated to silt and fine sand-sized particles. These aggregates were stable enough to withstand organic matter destruction and four hours of end-over-end shaking. This method of making parallel analyses of texture with and without dispersing agent, and thus of giving a quantitative empirical measure of stable microaggregation, is thought to be relevant to soil erodibility inasmuch as it reflects the stability of the microaggregates and also indicates a natural texture which can be expected to correlate better with porosity and rainfall acceptance than would true texture.

Table 5.9. Mechanical analysis of the first four horizons of a Kikuyu friable clay from Ruiru Coffee Research Station, Kenya, analysed with dispersing agent (above) and without dispersing agent (below)

Horizon depth (cm)	Sand 2000–600 μm	600–200 μm	200–60 μm	Silt 60–20 μm	20–6 μm	6–2 μm	Clay 0–2 μm	Texture
0–8	1·0	1·9	1·2	14·7	15·1	7·8	58·4	Clay
8–15	1·0	2·0	1·6	21·3	13·3	9·4	51·5	Clay
15–30	0·8	1·5	1·4	8·2	9·9	8·3	69·8	Clay
30–50	0·5	1·5	2·2	4·4	8·3	6·1	77·0	Clay
0–8	0·3	5·1	14·2	17·8	28·1	14·0	20·5	Silt loam
8–15	1·2	3·4	30·6	21·2	16·3	8·4	18·9	Loam
15–30	1·0	2·4	36·4	25·6	15·2	15·2	4·2	Silt loam
30–50	0·5	1·6	34·8	23·9	34·5	3·1	1·6	Silt loam

Another line of investigation concerns the use of rainfall simulators. Pereira (1956) subjected undisturbed soil cores 10 cm wide to artificial rain in the laboratory, using particularly large drops 6 mm in diameter which fell 2 m and reached 50 per cent of their terminal velocity. Exceptional intensities were needed to produce runoff in the soil samples used and Pereira reported little correlation between his findings and erodibility in actual practice.

The Kenya Soil Survey have recently carried out infiltration measurements with 10 cm rings and a 10 cm head of water in several areas. The variation between the 9 to 20 replicates used was sometimes large and the median reading was preferred to the arithmetic average. Infiltration using a rainfall simulator was tried out in the Kapenguria area (Braun and Kibe, 1975). Preliminary results suggest that high infiltration rates are found in sandy horizons of Arenosols and Chromic Luvisols, and in humous forest topsoils if not eroded. Vertisols are characterized by a high dry season infiltration but very low wet season infiltration. Low infiltration was found in some Calcaric Fluvisols and Gleyic Solonchaks. Infiltration measured by rainfall simulation was much slower than that determined with the 10 cm rings. In most cases infiltration increases caused by cultivation appear to be only very temporary, and infiltration is often greatly reduced by compaction caused by grazing animals, as has also been emphasized by Thomas (1974) in the Machakos area of Kenya and by Stephens (1971) in Uganda.

To measure soil erodibility by using plots, all other factors, including dimensions of the plot, plant cover or lack of cover, and gradient, need to be carefully standardized (Chapter 3.1). Such standardization, even on such an important and fundamental matter as the shape and dimensions of the plot, has been lacking in East Africa so far.

5.3.4. FARMING SYSTEMS IN EAST AFRICA IN RELATION TO EROSION HAZARD

Some general statements about the erosion hazards associated with different farming systems and types of land use appear to have gained general acceptance. It is often argued that shifting cultivation tends to reduce erosion hazards because crops are generally mixed, farm patches are small, and harvesting over a period coupled with lack of much weeding and maintenance gives some cover most of the time. More generally, it is widely accepted that perennial tree crops give greater protection than annuals. Although detailed studies in East Africa are few, some of those carried out have not always supported these conclusions, as when erosion plot studies in Tanzania at Tengeru, near Arusha, showed that clean weeded coffee lost far more soil than plots under maize (Anderson, 1962).

When farming systems are examined in more critical detail, the important factor appears to be the percentage cover provided, particularly at critical times of the year. Hudson (1971) has emphasized the importance of cover density, claiming that 'in the range of cover which is most usual for arable crops variations in the density of cover will influence erosion more than any other management factor'. His own work in Rhodesia appears to support this conclusion. He compared plots with maize at 36,000 and 24,000 plants per hectare and found erosion in the first only one-quarter of that in the second, and this he related to the fact that 10 per cent and 40 per cent of the soil was exposed in the two cases. Other experiments in Rhodesia (Hudson, 1957, 1959; Hudson and Jackson, 1959) found even minor differences in gradient associated with marked differences in soil losses. Protecting a plot with mosquito gauze reduced soil losses over six years to less than 1 per cent of those of an unprotected plot.

Some of the current evidence on farming systems in relation to erosion hazard is summarized below for each of the three territories, Kenya, Uganda and Tanzania. These summaries emphasize the great variety of climate and types of agriculture in the area, and tend to reinforce the view that the most important single factor affecting erosion is management—it is not so much the crop grown as the way it is grown—but that nevertheless there are also important intrinsic differences between soils which affect erosion hazard.

5.3.4.1. Kenya

About 70 per cent of Kenya is arid and described as range land. In these areas there is little agriculture of any sort. The highlands, in contrast, have some of the best agricultural land in Africa, though only about 17 per cent of the country has consistently over 600 mm of rain annually. In the highlands, increasing altitude is associated with higher rainfalls but also with decreasing temperatures, thus giving some vertical zonation and subtropical and temperate types of agriculture, and permanent, continuous farming rather than intermittent shifting agriculture is the rule, with arable annual crops (maize, beans, wheat), tree crops (tea, coffee) and good pasture. In the lower and warmer remaining densely populated areas around Lake Victoria and along the coast, agriculture is often based on natural fallow systems, but it is much modified by land shortages progressively forcing longer periods of cultivation. Some of the most difficult areas agriculturally are the intermediate areas of low and erratic rainfall where population pressures are nevertheless resulting in increasing annual arable cultivation and more and more overgrazing of the steepest land. The Machakos area discussed further below forms an example of such a situation.

In the highlands themselves, much steep land is cultivated intensively but erosion, though widespread, is not as severe as might have been expected due to (a) the excellent physical properties of many of the deep, red, porous and friable soils, developed over volcanic lavas and ashes, such as the Kikuyu friable clay discussed above, and (b) the fact that much bench terracing has been carried out in the past, and was at one time essential in order to secure registration of land above a certain gradient. Broad-based terraces (bunds) are recommended for slopes up to 4 per cent, narrow terraces, often with a V-shaped channel on the upper side, are recommended for intermediate slopes, and bench terraces for gradients of 20–30 per cent. According to Giçhungwa (1970), terrace spacing was based on the empirical formula:

$$\text{Vertical interval} = 0{\cdot}3\,\frac{(\%\ \text{slope} + 2)}{4}\ \text{metres}$$

but this gave terraces which were found to be too close. It is very doubtful if any single formula is generally applicable to all soils. Observation of erosion on the typical steep slopes of much of the highlands, often hand-planted to a mixture of maize and beans, suggests that farmers on the red soils can get away with farming practices which would spell disaster on less porous and stable soils. Working with such soils at Muguga, near Nairobi, Pereira *et al.* (1967) found that in a period of six years, only five storms led to substantial runoff.

In the heavy rainfall and generally steep Kericho area, also of red soils, a series of runoff plots planted to young tea was established in 1971 by Othieno (1973), using plots 18·3 m × 3·7 m with a slope of 10 per cent and the following treatments: bare soil, manually weeded; bare soil, chemically weeded; oats between tea rows; an *Eragrostis curvula* mulch. In general, both mulch and oats, once established, gave good protection, but bare soil, either hand-weeded or chemically weeded, lost about 11 per cent of total rain as runoff and 140 tonnes per ha of soil, equivalent to about 1 cm of topsoil. Soil loss was particularly rapid when the soil was already at field capacity prior to a storm. Othieno concluded that his results emphasized the importance of conservation when tea is young.

Earlier work carried out by EAAFRO and written up in a special number of the *East African Agriculture and Forestry Journal* (Pereira *et al.*, 1962) examined the hydrological effects of changes in land use in some East African catchment areas. Work on the development of tea estates in tall rain forest caused a considerable increase in storm runoff during the initial clearing and terracing operations, though the tea canopy once established gives good protection.

Using air photographs taken in 1948 and 1972, Thomas (1974) has recently completed a study of erosion in part of the Machakos area, where the average rainfall of 820 mm falls in two wet seasons. The steep slopes, with soil developed on the basement complex, are intensively used for livestock (cattle, goats, sheep) and for crops. The main crops, maize and beans, are planted at the beginning of one wet season and harvested before the beginning of the next; cultivation is by hand or with oxen, and sisal is grown along field boundaries and gully edges.

In 1948 the area was already severely gullied, some gullies reaching 10 m wide and 15 m deep, apparently formed during the preceding 30–40 years. Between 1955 and 1965 most cultivated land was terraced, usually with bench terraces. Results of this programme were spectacular, and some people (e.g. Kirby, 1968) claimed that erosion had been controlled.

Thomas, from his air photo analysis, concluded that between 1948 and 1972 gully extensions were small, with few new ones, but that sheet erosion had occurred. Cultivated land in the period increased by 44 per cent, but on the whole the cultivated land was better protected by banks than before, while the sisal plants and associated grasses they protect impeded soil and water movement downslope. Nevertheless, crop residues are not available as they are used for livestock feed or bedding, or are eaten by termites. Ploughing improves water retention and storage in fields of low gradient, but the bench terraces would be much more effective if a better vegetative cover were established on their banks. Because of terrace breaks, Thomas concludes that the finding of a suitable cover is the most immediate problem, but that the worst erosion is not in the cultivated areas, but in the steeper grazing areas which have been overgrazed, trampled and largely denuded.

The main problem in areas similar to the Machakos district appears to be that heavy rain at the beginning of each of two short rainy seasons falls when crops are still too young to give much cover. In Rhodesia, where there is a single, long wet season, Hudson (1971) has emphasized crop cover rather than mechanical measures which might be inadequately maintained and cause more damage if they break down than if they had never been installed at all, whereas in the drier parts of Kenya Thomas feels that crop cover is more difficult to achieve and that terraces thus become the main solution provided the steep banks can be protected with vegetation or with stones.

5.3.4.2. Uganda

The population of Uganda in the past was low enough in relation to land to be supported by stable natural rotation systems, but more recently population increases together with the introduction of cash crops have increased pressure on the land and shortened fallows markedly. In many areas fallows are down to about three years but preceded by a rotation of crops. Bananas, finger millet and cotton are the most important crops, with coffee, tea, and tobacco locally in evidence.

At Namulonge, soil losses in relation to rainfall intensity were studied for six years (Hutchinson *et al.*, 1958) using circular erosion plots $3m^2$ in area, but trials with local crops and farming systems do not appear to have been included. Percolation rate was preferred to runoff as an index of the effect of treatment since it is not affected by plot size. Predictably, both grass and grass mulch were associated with far less soil and water loss than were bare plots.

The extent of soil erosion in Uganda and current recommended practices have been summarized by Stephens (1971), who describes the general protection of arable land as being 'often good, especially in Kigezi and Teso districts'. On gentle slopes, grass wash stops across fields are recommended at about 30 m apart. On steeper slopes (southern Uganda and Mt Elgon), long bunds about 1 m wide and planted with *Paspalum notatum* are formed on the contour and tied at about 10 m intervals. On slopes over 20° in Kigezi, terraces about 10 m wide and separated by grass bunds or stone walls have been constructed.

Perennial crops in Uganda such as banana and coffee are generally mulched in most districts. Only sweet potatoes are planted on contour ridges. Cotton has been shown to yield better on the flat (Walton, 1962).

Serious erosion occurs on the steep cultivated slopes of the Ruwenzori and in the hilly, densely populated West Nile district. Landslips occur on the permeable soils of Mt Elgon. The coarse sandy loams of much of central and northern Uganda are particularly liable to rain-splash erosion when left unprotected. This can result in the formation of a loose sand topsoil which gullies easily, whereas on the more clayey soils of southern Uganda moderate sheet erosion occurs but gullies are less common.

Very heavy erosion is found in the east Karamoja district, now marked by widespread sheet erosion and spectacular gullies; there has been a change to a drier type of steppe vegetation within living memory. The basic cause here is overgrazing and soil compaction, leading to runoff and soil loss, and to progressively drier and poorer pastures until the stony subsoil is exposed. Dam construction in Karamoja is said to have made overgrazing even more serious by increasing cattle numbers. Farmers attach more importance to the quantity of their cattle than to their quality, so that Stephens sees the pastoral problem as 'primarily a social one'. He claims that in Uganda in general the worst examples of erosion are not on cultivated land but on communal grazing land, particularly where sandy soils are exposed to heavy rain and the dry season prevents the maintenance of a good vegetative cover. Rainfall acceptance is lowered by surface compaction, while in many hill areas indiscriminate burning encourages sheet erosion. Byelaws have attempted to control burning. A milder burn at the beginning of the dry season would cause less harm than a fierce burn at the end of it.

5.3.4.3. Tanzania

In Tanzania, as in Kenya, there are marked population concentrations in a few favoured areas which contrast with vast tracts of almost uninhabited land. However, there are few areas where rainfall is less than 500 mm, though rainfall distribution, with a generally long, very dry season from June to October, is unfavourable.

Coffee and bananas characterize the wetter, higher areas such as Kilimanjaro and Arusha where population densities are highest, and cotton and cattle the Sukamaland area south of Lake Victoria. The coastal belt produces maize, cassava, cashews and coconuts. In most of the rest of the country where crops are grown, maize is the chief food crop and sisal the main cash crop.

An excellent example of the disastrous effects of unadapted shifting cultivation on steep slopes is supplied by the recent agricultural history of the Uluguru mountains near Morogoro, central Tanzania. The rainfall of 1500–2850 mm supported mainly forest and woodland, until clearing began, probably about the year 1800, by Luguru tribesmen moving in from the Iringa area. They had been used to shifting cultivation on lowland plains and applied their methods with little adaptation to the steeper slopes of their new area. Emphasis now is on maize and hill rice. Cultivation is mainly by hoe and there is much burning. The Germans in 1909 created a forest reserve in part of the area, but this put more pressure on the remaining land, leading to shorter fallows and the spread of cultivation to steeper slopes. A classic vicious circle set in, with sheer erosion and lowered yields leading to more cultivation with less adequate fallows and thus to more sheet erosion. The net result was the progressive abandoning of some areas.

According to Temple (1972a), conservation policy in the Ulugurus has been marked by 'radical shifts in emphasis and direction'. In the 1930's emphasis was on ladder or step terraces. These are narrow terraces made with a hoe in which plant material and crop residues are spread out and covered with soil from the face of the terrace above. Live grass barriers and 1 m firebreaks round fields were encouraged, and in hilly areas regulations against all burning except when breaking new land were enforced periodically. Although excessive burning and sheet wash appear to have put considerable areas out of cultivation for long periods, these measures did have some good effect. In 1950, however, the Uluguru Land Usage Scheme, involving a different approach, began full operation (Young and Fosbrooke, 1960). Bench terracing on medium gradients was compulsory, and annual cropping forbidden on steeper slopes which were to be reforested. The bench terraces were, of course, much wider than the narrow ladder terraces and involved far more work and soil removal. No real information appears to have been available at the time on their effects on yields or on the dangers of landslides. From late 1952, each household was forced to construct 500 m^2 of bench terraces. Reduced burning was improving the cover and lessening stream flooding, but hostility to the scheme built up to a climax, and in mid-1955 all terracing stopped and large areas were set on fire in order to show defiance of the regulations. The scheme collapsed. Young and Fosbrooke (1960) say the scheme upset traditional social and cultural systems, but the technical recommendations were inadequate and discontent was exploited politically. Duff (quoted by Temple, 1972b) claims that the Luguru deliberately chose the poorer sites for terracing so as not to damage fertile land; on steep slopes broad terraces frequently exposed subsoil material.

At present it appears that landslides in the area are the most serious erosion threat (Temple and Rapp, 1972; Rapp *et al.*, 1972a, b). These are caused by a build-up of subsurface water pressure. Drainage is suggested as a countermeasure but is expensive even if technically feasible. Bench terracing appears to hold up water on slopes and to increase landslide danger. Even scattered trees help to control landslides by anchoring soil and removing some water, but this too impinges on customary practices, since the allocation of arable land is the privilege of clan heads, who resist taking land out of cultivation. Probably the only conservation method to find much local acceptance is the hoe-built ladder terrace.

A study of the Uluguru highlands was included in the important soil erosion research project carried out between 1968 and 1972 by the Universities of Dar es Salaam and Uppsala. Fifteen papers on erosion appeared in 1972 in a double volume of *Geografisker Annaler* (Rapp *et al.*, 1972a, b). These studies are concerned with two widely contrasted environments, mainly deforested mountain slopes and semi-arid interior plains with a long dry season such as occupy 60–70 per cent of the country (Peberdy, 1969).

Earlier EAAFRO work on water balances in catchment areas included studies in the Mbeya mountains (Pereira *et al.*, 1962), but only results for the first two years have been published so far and later work ran into recording difficulties and was abandoned in the late 1960s.

Trials in Tanzania on an erosion plot scale have been summarized recently by Temple (1972b).

At *Mpwapwa* plots with an area of 50 m^2 were set up on red sandy loam in the early 1930's (Staples, 1934, 1935, 1939) and included plots cultivated according to local practice to bulrush millet and to sorghum. The Mpwapwa rainfall of 690 mm falls mainly in the December–April wet season, with some intensive storms, mostly convectional; runoff from the cultivated plots was somewhat more than from grass, but very much less than from bare soil. After recording was restarted in 1946 (Rensburg, 1955), a plot under sorghum lost between 6 and 86 m^3/ha of soil per season over eight seasons as compared with only 0·3–0·9 m^3/ha from grass plots, and these losses were associated with a yield decline, but in another sorghum plot grass strips reduced losses by 53 per cent.

At *Lyamungu* plots 30·5 m long under coffee were established on deep, free-draining red soils in 1934. Ridges were found to have little effect unless closely spaced, but a banana mulch reduced soil loss to negligible amounts. Nevertheless, the coffee in the clean-weeded plots which lost the most soil was in the best condition because moisture retention four months after the rains was highest (Mitchell, 1965; Temple, 1972b).

At *Tengeru*, near Arusha, erosion plots on deep volcanic soil were established in 1954 and are of particular interest since they featured the local crops of coffee, maize and bananas. Soil losses from maize plots were considerably greater than from banana plots, but were reduced by closely spaced trash bunds. The trials showed that a permanent perennial crop of bananas can give a good cover and build up fertility. Bananas are associated with many steep East African mountain areas of dense population, where they form an important subsistence crop. Conversely, the high losses reported from clean-weeded coffee do not support the general idea that perennials always conserve soil better than annuals. Anderson (1962) made an interesting comparison between the soil losses and fertility, finding that the coffee plot which had lost most soil also had the lowest pH, and organic matter, total phosphorus and organic phosphorus contents.

The considerable differences between plot results at Mpwapwa, Tengeru and Lyamungu may be due to soil differences and to the greater incidence of runoff-producing storms in semi-arid central Tanzania as compared to the mountain footslopes in the north. At Mpwapwa millet and sorghum were planted in pure stands, but the local Gogo people prefer mixed cropping which probably gives a better cover and more protection, and the local habit of planting small fields also helps conserve soil, but if hoe-built contour ridges could be adopted, an excellent additional protection would be provided. The tying of ridges has been beneficial, particularly in dry years (Brown, 1963), but as pointed out by Russell (1973), this can be dangerous on soils subject to waterlogging. It is doubtful if tied ridging has as yet been adopted to any extent by local cultivators. With coffee the competition for water from cover crops has probably contributed to the fact that mulching is the favoured erosion control method, mostly with grass, though banana leaves are better and last longer. The Chagga and Meru farmers interplant coffee with bananas, thus lowering yields somewhat but greatly reducing water and soil losses, and this forms another example of local practice which has been justified by research.

REFERENCES

Ahn, P. M., 1968, The effects of large-scale mechanized agriculture on the physical properties of West African soils, *Ghana J. of Agricultural Science*, **3**, 93–97.

Ahn, P. M., In preparation, Quantitative determinations of the micro-aggregation of the Kikuyu friable clays of the Kenya highlands, University of Nairobi, Dept. of Soil Science, Technical Communication 7.

Anderson, G. D., 1962, The effect of various cropping systems on soil fertility, Paper read at the East African Soil Fertility Specialist Committee Meeting, Tengeru (unpublished).

Bouyoucos, G. J., 1935, The clay ratio as a criterion of susceptibility of soils to erosion, *J. Amer. Soc. Agron.*, **27**, 738–741.

Braun, H. M., In preparation, Seasonal rainfall probability for three areas in Kenya (Embu-Kibwezi, Kapenguria-Kakamega, Coast) with some applications to agricultural problems, Kenya Soil Survey, Ministry of Agric., Nairobi, Misc. Report No. 1.

Braun, H. M., and J. M. Kibe, 1975, Infiltration measurements, In *Kenya Soil Survey, Annual Report for 1974*, National Agricultural Laboratories, Kenya Ministry of Agriculture, Nairobi.

Brown, K. J., 1963, Rainfall, tie-ridging and crop yields in Sukumaland, Tanganyika, *Empire Cotton Growing Review*, **40**, 34–40.

Brown, L. H., and J. Cocheme, 1969, Technical report on a study of the agroclimatology of the highlands of eastern Africa, FAO, Rome.

Bryan, R. B., 1968, The development, use and efficiency of indices of soil erodibility, *Geoderma*, **2**, 5–26.

Gichungwa, J. K., 1970, *Soil Conservation in Central Province* (*Kenya*), Ministry of Agriculture, Nairobi.

Hudson, N. W., 1957, Erosion control research–progress report on experiments at Henderson Research Station 1953–1956, *Rhodesian Agric. J.*, **54**, 297–323.

Hudson, N. W., 1959, Results of erosion research in Southern Rhodesia, Advisory leaflet 13, Fed. Dept. of Conservation and Extension, Salisbury.

Hudson, N. W., 1965, The influence of rainfall on the mechanics of soil erosion, M.Sc. Thesis, University of Cape Town.

Hudson, N. W., 1971, *Soil Conservation*, Batsford, London.

Hudson, N. W., and D. C. Jackson, 1959, Results achieved in the measurement of erosion and runoff in Southern Rhodesia, *Proc. Third Inter-African Soils Conf.*, Dalaba, 573–583.

Hutchinson, J., H. L. Manning and H. G. Farbrother, 1958, On the characterisation of tropical rainstorms in relation to runoff and percolation, *Quarterly J. Roy. Met. Soc.*, **84**, 250–258.

Keter, J. K. A., 1974, A study of soil phosphorus forms in Kikuyu red loam and of their surface activity, M.Sc. Thesis, University of Nairobi, Dept. of Soil Science.

Kirby, C. P., 1968, *East Africa*, Benn, London.

Lawes, E. F., 1969, Some confidence limits of expected rainfall, E. African Met. Dept., Nairobi, Technical Memorandum No. 15.

Lumb, F. E., 1971, Probable maximum precipitation in East Africa for durations up to 24 hours, E. African Met. Dept., Nairobi, Technical Memorandum No. 16.

Mitchell, H. W., 1965, Soil erosion losses in coffee, *Tanganyika Coffee News*, April/June, 135–155.

Othieno, C. O., 1973, First year studies on rainfall, runoff and soil erosion in a field of young tea, Paper read at the 5th Specialist Committee on Applied Meteorology, Nairobi, Sept. 1973 (unpublished).

Ovens, G., 1969, A study of soil erodibility by separation of the primary stages of the erosion process, M.Sc. Thesis, University of Reading.

Peberdy, J. R., 1969, Rangeland, In W. T. W. Morgan (Ed.), *East Africa: Its Peoples and Resources*, Oxford University Press, Nairobi.

Pereira, H. C., 1956, A rainfall test for structure in tropical soils, *J. Soil Sci.*, **7**, 68–74.

Pereira, H. C., P. H. Hosegood and M. Dagg, 1967, Effects of tied ridges, terraces and grass leys on a lateritic soil in Kenya, *Exp. Agric.*, **3**, 89–98.

Pereira, H. C., J. S. G. McCulloch, M. Dagg, O. Kerfoot, P. H. Hosegood and M. A. C. Pratt, 1962, Hydrological effects of changes in land-use in some East African catchment areas, *E. Afr. Agric. For. J.*, **27** (Special Issue), 42–75.

Rapp, A., V. Axelsson, L. Berry and D. H. Murray-Rust, 1972, Soil erosion and sediment transport in the Morogoro river catchment, Tanzania, *Geografiska Annaler*, **54A**, 3–4, 125–155.

Rapp, A., L. Berry and P. H. Temple (Eds.), 1972a, Studies of soil erosion and sedimentation in Tanzania, *Geografiska Annaler*, **54**, 3–4, 105–379.

Rapp, A., L. Berry and P. H. Temple, 1972b, Soil erosion and sedimentation in Tanzania—the project, *Geografiska Annaler*, **54**, 3–4, 105–109.

Rensburg, H. J. van, 1955, Runoff and soil erosion tests, Mpwapwa, Central Tanganyika, *E. Af. Agric. J.*, **20**, 228–231.

Russell, E. W., 1973, *Soil Conditions and Plant Growth*, 10th edn., Longmans, London.

Staples, R. R., 1934, 1935, 1939, Reports of the runoff and soil erosion tests at Mpwapwa in semi-arid Tanganyika, *Annual Reports*, Department of Vet. Science, Animal Husbandry, Tanganyika, 1933, 1934 and 1939.

Stephens, D., 1971, Soil fertility, In J. D. Jameson (Ed.), *Agriculture in Uganda*, Oxford University Press.

Taylor, C. M. and E. F. Lawes, 1971, Rainfall intensity—duration–frequency data for stations in East Africa, East African Met. Dept. Nairobi, Technical Memorandum No. 17.

Temple, P. H., 1972a, Soil and water conservation policies in the Uluguru mountains, Tanzania, *Geografiska Annaler*, **54A**, 3–4, 110–123.

Temple, P. H., 1972b, Measurements of runoff and soil erosion at an erosion plot scale with particular reference to Tanzania, *Geografiska Annaler*, **54A**, 3–4, 203–220.

Temple, P. H., and A. Rapp, 1972, Landslides in the Mgeta area, western Uluguru mountains; geomorphological effects of sudden heavy rainfall, *Geografiska Annaler*, **54A**, 3–4, 157–193.

Walton, P. D., 1962, Cotton and agronomy trials in northern and eastern provinces of Uganda 1956 to 1961, *Cotton Growers Review*, **39**, 114–24.

Young, R., and H. Fosbrooke, 1960, *Land politics among the Luguru of Tanganyika*, Routledge & Kegan Paul, London.

5.4

Application of the Universal Soil Loss Equation of Wischmeier and Smith in West Africa

E. J. Roose

5.4.1. INTRODUCTION

The Universal Soil Loss Equation (USLE) of Wischmeier and Smith,:

$$A = RKLSCP, \tag{5.1}$$

has been discussed in earlier chapters.

Knowing the climatic aggressivity of a region, the erodibility of the soil and the topography of the land to be cultivated, the equation can be used to determine the anti-erosive techniques required to ensure under a given cropping practice that soil losses do not exceed a certain tolerable limit, usually taken to be 2 to 12 tonnes/ha/yr depending on the soil characteristics.

The purpose of this report is to examine the applicability of the USLE for West Africa. These results are based on experiments conducted at l'Office de la Recherche Scientifique et Technique Outre-Mer (ORSTOM) and the Instituts Francais de Recherches Appliquees.

5.4.1.1. The Rainfall Factor (*R*)

The regression analyses between rainfall and runoff and soil loss on small bare fallow plots at Adiopodoumé reported by Roose (1973b) indicate that the amount of rainfall alone does not explain the erosion phenomenon adequately. Soil moisture content prior to the rain and the rainfall intensity are also involved. The correlation between rainfall and erosion becomes significant only for maximum intensities, measured over more than 20 minutes for soil loss and over 10 minutes for surface runoff.

In contrast with observations from the Mediterranean and Saharan zones (Heusch, 1970; Roose, 1971, 1972a), where the decennary and centenary exceptional rains transform the landscape radically, but in accordance with Wischmeier and Smith (1958, 1960), the level of erosion in the humid tropical environment is not determined by the exceptional rain but by the sum of the 10 or 20 most erosive rainstorms of the year. In their rainfall factor, calculated on a diurnal, monthly or annual basis, Wischmeier and Smith (1958, 1960) used the combined effect of rainfall amount, maximum intensity and duration of the individual rainstorms. The results showed that there were large variations of the erosivity (R) factor for rains of the same amount and that the variation increased rapidly with the amount. The problem of developing appropriate erosivity indices was discussed more fully in Chapter 3.1.

Many researchers (e.g. Charreau, 1969; and Chapter 3.2) have observed linear relationships between kinetic energy (E) of rainfall and its amount, but the relation between the maximum intensity occurring within the initial 30 minutes of the rain (I_{30}) and the total amount is curvilinear, with a low correlation coefficient (Charreau, 1969). Since the Wischmeier R factor is the product of the kinetic energy at the maximum 30-minute intensity, a multiplication of errors occurs when attempts are made to derive the EI_{30} index from data for rainfall amounts. Nevertheless, Delwaulle (1973) and other French workers have developed a regression equation to estimate R from the rainfall amount (H) and its maximum intensity within 30 minutes (I_{30}). The regression was developed for Upper Volta and Niger but may be applicable elsewhere in West Africa. The equation is:

$$R = 0{\cdot}0157H \times I_{30} - 1{\cdot}18. \quad (5.2)$$

Computation of I_{30} from rainfall intensity records in West Africa is difficult because such data are not available for many locations. However, a large number of weather stations in West Africa do have daily rainfall records for the past 20 to 50 years. Variation in EI_{30} with rainfall amount can be readily calculated for stations where intensity is recorded (Table 5.10) and extrapolated to the large number of stations where amounts are available.

The analysis of the daily rainfall amount, as measured at 8 am every day, with the EI_{30} index calculated from the few existing recording raingauges, indicates the following:

1. In the region of the Ivory Coast within 40 km of the sea there is a rectilinear regression ($R = 0{\cdot}577$, $H = 5{\cdot}77$) between R and H for the 'monsoon' rains from June to September and a curvilinear regression for the rains in the rest of the year.
2. This curvilinear regression $\log R = f(\log H)$ is the same for all the stations investigated, viz., Abidjan, Amaguie, Divo, Bouake, Dorchogo in Ivory Coast, Niangoloko, Koudougou and Ouagadougou in Upper Volta.
3. The computation of R from daily rainfall amounts indicates that the mean difference over five years with respect to the measured annual factor R did not exceed 5 per cent. Therefore one can transform the observations of diurnal rainfalls (20 to 50 years) into erosivity factors (R) and prepare a tentative isoerodent map (Figure 5.1).
4. There exists a simple relation between the mean annual factor (R) over a 5 to 10 year period, and the mean annual rain (H) during the same period (Table 5.10).

$$R/H = 0{\cdot}5 \pm 0{\cdot}05 \quad (5.3)$$

Table 5.10. Variability in EI_{30} with amount of rainfall for stations in West Africa. From 'Erosion et ruissellement en Afrique de l'Ouest—Vingt années de mesure, en petites parcelles expérimentales', ORSTOM/IITA. Reproduced by permission of F. Bonnet-Dupeyron, ORSTOM

Rainfall	Abidjan		Divo		Bouake		Korhogo		Saria/Gonse		Allokoto	
(mm)	Variation	Mean	Variation	Mean	Variation	Mean	Variation	Mean	Variation	Mean	Variation	Mean
15	0·5–8	5	1–7	4·5	0·5–7	3·7	1·5–7	5	2·3–6·5	4·7	1·5–8	5
30	6–24	18	7–30	17·6	4–22	16	4–26	17	16–31	18	6–20	18
50	12–66	44	20–90	48	15–80	40	19–80	50	30–65	48	30–65	50
70	32–90	78	49–108	84	25–100	73	31–120	83	84–104	86	—	—

This ratio has been found to hold for a number of locations in the Ivory Coast, Upper Volta, Senegal, Niger, Tchad, Cameroun, and Malagasy, with the exception of stations situated near mountains (Dschang and Befandriana) and at Bouake (Table 5.10).

According to Wischmeier, for the American plains at least 22 years of observations were needed to develop an adequate set of data. Thus it may be preferable for the present to use the ratio 0·5 rather than the measured ratio for 10 years or less of observations.

The tentative erosivity map for West Africa presented here (Figure 5.1) is an initial step pending the availability of more detailed rainfall records from a large number of stations. Apart from the hilly regions and coastal plains, the accuracy of the estimated climatic factor is thought to be sufficient for it to be used in the USLE on a tentative basis.

It is evident from this map that R is very high for the more humid areas and decreases with the decrease in rainfall at the coast (Abidjan, $H = 2100$ mm; $R = 1260$) and Ouagadougou ($H = 830$ mm; $R = 430$). It is interesting to compare some mean annual erosivity factors (R) for various regions with the tentative values for West Africa:

60 to 300 in Tunisia (Masson, 1971)
50 to 300 in Morocco (Kalman, unpublished)
60 to 340 in the middle of France (Masson, 1971)
150 to 650 in the USA (Wischmeier, 1960)
200 to 600 in Upper Volta (Roose, Arrivetz and Poulain, ORSTOM IRAT unpublished report, 1974)
500 to 1400 in Ivory Coast (Roose, 1973b)

Table 5.11. Vegetal cover factors (C) for some West African conditions. From 'Erosion et ruissellement en Afrique de l'Ouest—Vingt années de mesure, en petites parcelles expérimentales', ORSTOM/IITA. Reproduced by permission of F. Bonnet-Dupeyron, ORSTOM

	C, representative annual value
Bare soil	1·0
Dense forest or culture with a thick straw mulch	0·001
Savannah and grassland, ungrazed	0·01
Cover crops: late planted or with slow development	
First year	0·3–0·8
Second year	0·1
Cover crops with rapid development	0·1
Maize, sorghum, millet	0·3–0·9
Rice (intensive culture, second cycle)	0·1–0·2
Cotton, tobacco (second cycle)	0·5
Groundnuts	0·4–0·8
Cassava (first year)	0·2–0·8
Palms, coffee, cocoa, with cover crops	0·1–0·3

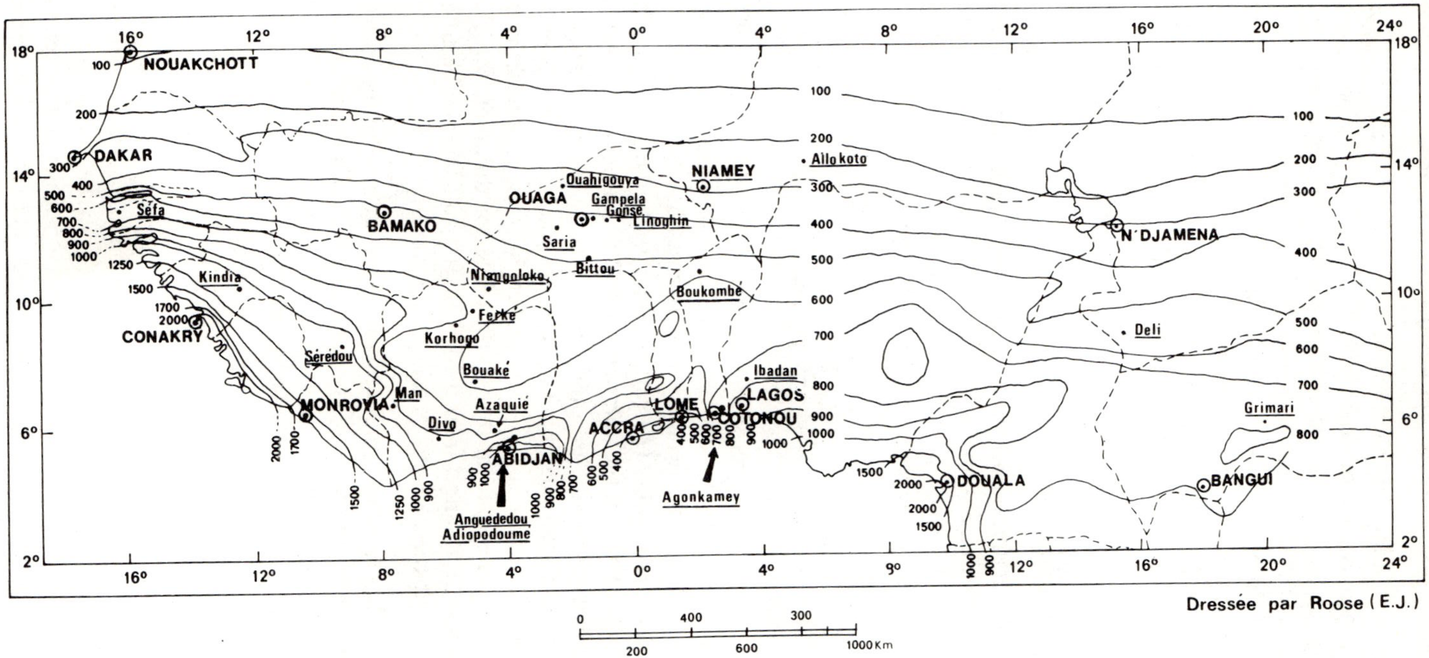

Figure 5.1. Tentative erosivity map (R) for West Africa and location of runoff and erosion plots. From rainfall data of Service Hydrologique de l'ORSTOM, compiled by E. J. Roose, 1975. Reproduced by permission of F. Bonnet-Dupeyron, ORSTOM

5.4.1.2. Vegetal Cover and Cultural Techniques: Factor *C*

Vegetal cover is an important factor in erosion control (Table 5.11). Burning in the savannah increases runoff and its velocity, and unpublished results of the author show that the intensity of erosion can then be extremely high.

The *C* factor of the Wischmeier equation is the ratio between the measured soil loss under a certain management and the soil loss observed on a standard bare plot. It expresses the interaction of the crop and the cultural techniques. Using a single global value to approximate the various values found during the year, the annual values in West Africa (Roose, 1973b) are given in Table 5.11.

5.4.1.3. Soil Erodibility Factor *K*

To determine the relative erodibility of different soil types, Wischmeier proposed comparisons of erosion from standard plots, 22·2 m long, 9 per cent slope and kept bare. The plots were superficially tilled to avoid crust formation. By choosing characteristic slopes and correcting the results for the topographic factor (*SL*), the soil erodibility *K* may be calculated as

$$K = \frac{A}{R \times LS \times 2{\cdot}24} \tag{5.4}$$

where A is soil loss in tonnes/ha/yr, R the rainfall factor, LS the landform factor and 2·24 the coefficient to convert the English units (ton/acre) used in Wischmeier's equation into decimal units (tonnes/ha).

In the USA, the erodibility (K) varies from 0·03 to 0·6 for a majority of soils (Wischmeier *et al.*, 1971). If the nomograph of Wischmeier *et al.* (1969) is used to estimate erodibility using silt, fine sand (20 to 100 μm), organic carbon content, a structure factor and a permeability factor, it is found that for '*sols ferrallitiques*' on tertiary sand, *K* ranges from 0·05–0·10, for those on granite, from 0·10–0·15, and for those on schists, from 0·15–0·18, while for different '*sols ferrugineux tropicaux*' on granite, it ranges from 0·20–0·30. It would be valuable to have more field plot data, but these results at least suggest:

1. The variability in *K* over time necessitates replications of the determinations in both space and time because of the residual effect of organic matter and the timing of rain and land preparation. In practice, *K* should be monitored on plots which have been cultivated for at least three or four years, the crop residue being removed and two to three harrowings made each year to break any crusts developed. Alternatively, to maintain a clean soil without tillage, chemical herbicides can be used.

2. The '*sols ferrallitiques*' are very resistant to erosion and the 'sols ferrugineux tropicaux' are moderately resistant after two or three years' cultivation. The greater susceptibility of the '*sols ferrugineux*' can be explained by the low permeability of these soils, their tendency to develop crusts and the relatively high percentage of silt and fine sand and low organic matter content. Nevertheless,

they are often more resistant to erosion than some Alfisols of the temperate region. In Africa, high erosivity seems to be more the cause of excessive erosion than high soil erodibility.

3. Measured erodibility values (Roose, 1973b) and those obtained using the nomograph compare reasonably well for the nine soils studied, with the exception of gravelly soils. The nomograph does not apply to soils with a high gravel content in the surface layer. Gravel and rock fragments often act as a mulch in the tropics (Dumas, 1965). At Korhogo, for example, not only is K very low (0·02) but it decreases with an increase of gravel in the surface horizon.

Although the soil loss equation seems to be applicable to the soils with kaolinite as dominant clay mineral, it may or may not apply to soils with expanding lattice clay minerals, for example vertisols. The experiments of Heusch (1970) in Morocco show that vertisols react very differently.

5.4.1.4. The Landform Factor *LS*

Though soils with slopes of less than 2 per cent erode (Fournier, 1967), soil losses generally increase exponentially with the steepness of the slope (Chapter 4.3). Wischmeier and Smith (1960) found that an equation of the second degree was better than a logarithmic function.

The influence of the *length of the slope* is not so important as degree of slope. Wischmeier *et al.* (1958) concluded that the relations between soil loss and the length of the slope can vary from soil to soil and from year to year. The magnitude of the exponent (0·1–0·9) is a function of the soil characteristics, vegetal cover, crop residues, and other factors. Wischmeier and his collaborators at the University of Purdue recommended that the LS factor be determined from the equation

$$LS = \frac{\sqrt{L}}{100}(0{\cdot}76 + 0{\cdot}53S + 0{\cdot}076S^2) \qquad (5.5)$$

where L is the length of the slope in feet and S the slope percentage. For a plot with 9 per cent slope and 72·6 feet long, the LS factor is 1.

At Adiopodoume, on bare soil over five years, similar results to those of Wischmeier and Smith were obtained, although they were very variable from one year to another (Roose, 1973b). Small variations in slope (0·5 per cent) are sufficient to produce notable variations in soil loss and sometimes in runoff (Table 5.12) (Roose, 1967).

The LS factor, and its relation with erodibility, is not very satisfactory, because it is evident that the effect of slope length and angle on soil loss are not independent of the R, C and K factors. Until the relation between the slope length and soil loss is better defined, construction of terraces and other mechanical control measures will present difficult design problems.

Table 5.12. Influence of slope on runoff (% of rainfall) and erosion. From 'Erosion et ruissellement en Afrique de l'Ouest—Vingt années de mesure en petites parcells expérimentales', ORSTOM/IITA. Reproduced by permission of F. Bonnet-Dupeyron, ORSTOM

Adiopodoume (Ivory Coast) 1968–1972	Slope (%) 4·5	7	20
Erosion (tonnes/ha/yr)	60	138	570
Mean runoff (%)	35	33	24
Max. runoff (%)	74	74	68
Sefa (Senegal) 1955–1968	1·25	1·50	2
Mean erosion (tonnes/ha/yr)	5	8·6	12
Mean annual runoff (%)	16	22	30

5.4.1.5. Erosion Control Measures: The *P* Factor

Anti-erosive techniques such as buffer strips, contour bunds or mulching can be very important for erosion control. When grass strips are used the well-developed root systems help to maintain a high permeability, so that the strips absorb much runoff. They also intercept runoff and so help deposit some of the eroded soil (Roose and Bertrand, 1971). At Adiopodoume in the Ivory Coast, over a period of five years, there were natural ridges as high as 50 cm formed from the soil deposited against the grass strips.

A straw mulch of some centimetres of thickness absorbs the kinetic energy of the rains as efficiently as a secondary forest and keeps the soil loss (0·03 tonnes/ha/yr) and runoff (0·5 per cent) within acceptable limits (Table 5.13).

Soil conditioners can also play an important role. Spraying a thin permeable film of a polyvinyl alcohol (Curasol, Hoechst) at the rate of 60 g/l/m^2 after ploughing reduced soil loss by 40 to 90 per cent and runoff by 25 to 55 per cent in

Table 5.13. Representative values of the erosion control factor *P*. From 'Erosion et ruissellement en Afrique de l'Ouest—Vingt années de mesure, en petites parcelles expérimentales', ORSTOM/IITA. Reproduced by permission of F. Bonnet-Dupeyron, ORSTOM

USA	(Wischmeier and Smith, 1965)	*P*
	Contour ploughing	0·75
	Ploughing and contour trenches	0·5
	Ploughing and contour grass strips	0·25
West Africa	(Roose, 1973a, b; Delwaulle, 1973)	
	Contour trenches	0·20–0·10
	Anti-erosive strips (2–4 m wide)	0·30–0·10
	Straw mulch (6 tonnes/ha)	0·01
	Curasol (60 g/l/m^2)	0·50–0·20
	Temporary grassland	0·5–0·1

comparison with the control. The use of polymers for erosion control can be highly effective, but at present is not an economic proposition for arable agriculture (Chapter 4.4).

Biological methods, involving maximum cover of the soil, mulch, cover crops, mixed and relay crops, are more efficient erosion control methods than mechanical designs (terracing, etc.). The mechanical designs are also expensive and difficult to maintain. However, it is the engineering practices that are more frequently described in soil conservation manuals, and often recommended without studies to determine whether they provide an effective and economic method of control, adapted to the ecological and farming conditions of the region.

5.4.2. CONCLUSIONS

(i) The application of the Universal Soil Loss Equation in the humid parts of West Africa can predict sheet and rill erosion in zones away from mountains.

(ii) Well-designed replicated experiments are needed, continued for several years at several sites, to obtain all the parameters in the Universal Soil Loss Equation.

(iii) The data obtained from small plots may pose some problems, especially if the results are to be extrapolated to large catchments. The equation neglects the qualitative aspects of the eroded materials.

Many tropical soils have their nutrient supplies concentrated in the upper 20 cm, especially of the forest zone, and sheet erosion takes selectively the organic and mineral colloids and the nutrient elements (Roose, 1967, 1973b; Lal, 1976).

The USLE may not be applicable to vertisols and soils with expanding lattice clay minerals. The topographic factor, and especially the length of the slope, is the weak point in this equation because its effect varies with soil type, texture and vegetal cover.

If the erosivity, erodibility, land form and management factors are determined, the equation can predict soil loss for the majority of cultivated land in West Africa, and especially for gentle and moderate slopes, covered by '*sols ferrallitiques et ferrugineux tropicaux*'.

One question that remains for consideration is whether erosion control measures should be based on mean annual erosivity, or on the exceptionally destructive storm, likely to occur once in ten, or perhaps one hundred years.

SUMMARY

Based on more than 400 plot-year data from erosion plots of nine stations in five West African countries, a critique is made of the Universal Soil Loss Equation in relation to its use in West Africa. Preliminary results from this region suggest that there exists a simple approximate relationship between the annual climatic erosion index $R\,(EI_{30})$ and the mean annual rainfall (H), namely $R = \frac{1}{2}H$, making it possible to extrapolate the results to compute R. The erosivity increases from the Sahel ($R = 100$) to the most humid tropical zones, where extremely high values are encountered ($R = 1000$ to 2000).

The vegetal cover is the most important factor in preventing runoff and erosion, regardless of climatic aggressivity, the slope and the soil type. However, cultural techniques can significantly influence the erosion when the soil is partly covered.

The resistance of the soil can be approximately evaluated using the nomograph of Wischmeier, with appropriate adjustment for the gravel contents of the soils, but further evaluation of the topography factors (L and S) is needed.

Erosion control techniques (P) involving terracing are both expensive and less efficient than biological control methods, which establish and maintain a vegetal cover by thick and early sowing, correct cultivations, adequate and balanced fertilization, mulching with straw, use of rotations, and buffer strips.

Although the Wischmeier equation enables the factors determining erosion in territories in West Africa to be defined, it is not universal because it is not adapted to regions where gully erosion is dominant, as in the more mountainous zones.

REFERENCES

Bertrand, R., 1967, Etude de l'érosion hydrique et de la conservation des eaux et du sol en pays Baoulé, *Coll. sur la Fertilité des Sols Tropicaux, Tananarive*, 19–25/11/67, No. 106, pp. 1281–1295.

Charreau, C., 1969, Influence des techniques culturales sur le développement du ruissellement et de l'érosion en Casamance, *Agron. Trop. Paris*, **24**, 836–842.

Charreau, C., and R. Nicou, 1970, L'amélioration du profil cultural dans les sols sableux et sablo-argileux de la zone tropicale sèche Ouest-Africaine et ses incidences agronomiques, *Agron. Trop. Paris*, **26**, 209, 565, 903, 1183.

Cormary, Y., and J. Masson, 1964, Etude de conservation des eaux et du sol au Centre du Génie Rural de Tunisie. Application à un projet type de la formule de perte de sols de Wischmeier, Cah. ORSTOM, série Pédol., v. **2**, No. 3, pp. 3–26.

Delwaulle, J. C., 1973, Résultats de six années d'observations sur l'érosion au Niger, *Bois et Forêts des Tropiques*, **150**, 15–37.

Dumas, J., 1965, Relation entre l'érodibilité des sols et leurs caractéristiques analytiques, Cah. ORSTOM, sér. Pédol., **3**, 4, 307–333.

Fournier, F., 1962, Carte du danger d'érosion en Afrique au Sud du Sahara fondé sur l'agressivité climatique et la topographie, Notice explicative, CEE–CCTA, Paris.

Fournier, F., 1967, La recherche en érosion et conservation des sols sur le continent Africain, *Sols Africains*, **12**, 1, 5–53.

Goujon, P., 1968, Conservation des sols en Afrique et à Madagascar. 1ère Partie: les facteurs de l'érosion et l'équation universelle de Wischmeier, *Bois et Forêts des Tropiques*, **118**, 3–17.

Heusch, B., 1970, L'érosion du pré-Rif. Une étude quantitative de l'érosion hydraulique dans les collines marneuses du Pré-Rif occidental, In *Annales de la Recherche Forestière au Maroc*, Numéro spécial, Etudes sur l'érosion, **12**, 9–176.

Hudson, N. W., 1963, An historical background of erosion and its control, *Rhodesia Agric. Journal.*, Nov/Dec, Bulletin No. 2226.

Jackson, S. P., 1961, *Atlas Climatique de l'Afrique*, CCTA/CSA, Lagos-Nairobi.

Lal, R., 1976, *Soil erosion on Alfisols in Western Nigeria. I: Effect of Slope Rotation and Residue Management*, Geoderma, **16**, 363–375.

Masson, J. M., 1971, L'érosion des sols par l'eau en climat méditerranéen. Méthode expérimentale pour l'étude des quantités de terre érodée à l'échelle du champ, Thèse Doct. Ing. Univ. Sciences et Techniques du Languedoc. CNRS No. AO 5445.

Neal, J. G., 1938, The effect of the degree of slope and rainfall characteristics on runoff and soil erosion, *Missouri Agr. Expt. Sta. Res. Bull.* 280.

Roose, E. J., 1967, Dix années de mesure de l'érosion et du ruissellement au Sénégal, *Agron. Trop.*, **22**, 2, 123–152.

Roose, E. J., 1971, Note technique concernant l'érosion hydrique au Maroc, *Bull. de liaison des Ing. Forestiers du Maroc*, No. 6, 47–52.

Roose, E. J., 1972a, Comparaison des causes de l'érosion et des principes de lutte antiérosive en région tropicale humide, tropicale sèche et méditerranéenne, Communication aux *Journées d'Etude du Génie Rural à Florence* du 12 au 16/9/72, pp. 417–441.

Roose, E. J., 1973a, Natural mulch or a chemical soil conditioner for reducing soil erosion in humid tropical areas, In *Soil Conditioners*, Soil Sci. Soc. Amer., Spec. Pub., 131–138.

Roose, E. J., 1973b, Dix-sept années de mesures expérimentales de l'érosion et du ruissellement sur un sol ferrallitique sableux de basse Côte d'Ivoire. Contribution à l'étude de l'érosion hydrique en milieu intertropical, ORSTOM, Abidjan, Thèse Doct. Ing., Fac. Sci. Abidjan, 1973.

Roose, E. J., and R. Bertrand, 1971, Contribution à l'étude de la méthode des bandes d'arrêt pour lutter contre l'érosion hydrique en Afrique de l'Ouest, Résultats expérimentaux et observations sur le terrain, *Agron. Trop. Paris*, **26**, 1270–1283.

Wischmeier, W. H., and D. D. Smith, 1958, Rainfall energy and its relationship to soil loss, *Trans. Am. Geophys. Union*, **39**, 285–291.

Wischmeier, W. H., and D. D. Smith, 1960, A universal soil loss equation to guide conservation farm planning, *Trans. 7th Int. Congr. Soil Sci.*, **1**, 418–425.

Wischmeier, W. H., D. D. Smith and R. E. Uhland, 1958, Evaluation of factors in the soil loss equation, *Agric. Eng.*, **39**, 458–462.

Wischmeier, W. H., C. B. Johnson and B. C. Cross, 1969, A soil erodibility nomograph for farmland and construction sites, *J. Soil Water Conservation*, **26**, 189–193.

Sources of Erosion Data.
Stations in Francophone West Africa used for recording erosion

Country	Location	No. of plots	Slope (%)	Years of study	Rainfall (mm/an)
Ivory Coast	Adiopodoumé	9	4, 7, 22	1956–74	2200
	Bouaté	5	4	1960–74	1200
	Korhogo	3	3	1965–75	1400
Upper Volta	Gampela	4	0·7	1966–71	800
	Gonsé	1	0·5	1968–74	800
	Saria	4	0·7	1971–74	830
Bénin	Agonkamey	4	4	1965–71	1200
Senegal	Séfa	10	1 to 2	1954–68	1400
Niger	Allokoto	4	3	1966–71	500

5.5

Erosion and Mechanization

R. FAUCK

The spread of mechanization of agriculture in humid tropical Africa has been characterized by a rapid increase in soil erosion. This fact has brought many to recommend, if not the rejection of mechanization, great prudence in its systematic utilization. A careful appraisal leads to the conclusion that mechanization should be examined critically in relation to existing conditions, particularly the various types of soils where it is to be applied.

5.5.1. THE MODALITIES OF EMPLOYING MECHANIZATION

All large mechanized cultivation projects in forest regions start by clearing the trees, mostly using powerful machines, followed by systematic removal of roots and stumps. The entire operation involves felling, uprooting, root-cuttage and soil inversion before planting the crop. Planting presupposes a certain amount of tillage. Tillage operations are done with three objectives. The first is the preparation of the soils by ploughing, inversion and pulverization in order to create a seedbed. This means creating an environment where the seeds find suitable conditions to germinate. The second objective is to improve the infiltration of rain. The last involves the post-planting operations such as interculture.

All these operations have different influences on soil erosion, but it is difficult to separate the individual effects since they all interact. When other factors remain constant, erosion is a function of soil depth, water content and time of rainfall in relation to tillage operations. In addition, the results vary between sandy types (12 per cent of clay in Casamance) and those of high clay content (65 per cent clay in the valley of Niari in the Congo). It is difficult to separate the variables and to define, among the mechanization methods, those factors which increase erosion. It is the complexity of these factors which has resulted in controversy over mechanization.

5.5.2. THE PRINCIPAL RESULTS

Some researchers have measured directly the influence of mechanization on the amount of soil loss and runoff. There are other experiments which deal with the evolution of the physicochemical characteristics of mechanically cultivated soils (Cointepas, 1958; Cunningham, 1963; Dumas, 1965; Pereira *et al.*, 1967; Goujon *et al.*, 1968; Seguy, 1969; Fauck *et al.*, 1969; Charreau and Seguy, 1969; Le Buanec, 1971; Roose, 1973a; Lal, 1974).

Most data come from experiments conducted in Senegal (Sefa, Bambey), in the Ivory Coast (Adiopodoume, Bouake), in Upper Volta (Saria), in Benin (Cotonou) and in Nigeria. The following conclusions can be drawn.

Clearing disturbs soil the most, resulting in an heterogeneous A horizon, bringing gravel to the surface, causing rapid changes in the chemical and physical characteristics, notably a decrease in organic matter and nitrogen contents and structural stability, and a spectacular increase in soil loss and runoff. But it is difficult to attribute the soil erosion to mechanization, since forest removal and other operations also disturb the ecological equilibrium. Nevertheless, excessive tillage and the heterogeneity resulting from uprooting are certainly responsible for a decrease in the physical stability of the surface horizons and an increase in the susceptibility of the soil to erosion. In this connection, it is advisable to attempt to improve the soil structure destroyed by clearing and uprooting by growing leguminous or other cover crops before adopting arable cultivation.

The conventional method of seedbed preparation consists of ploughing, pulverizing, rolling, harrowing, spreading of fertilizers and sowing. The effects of these operations should be carefully analysed, because the results obtained are sometimes contradictory. The different results can be explained not only by the diversity of the techniques employed, but also by the influence of the soil type and the variability of the clay and silt content in the upper horizons. For instance, fairly sandy soils, such as those in Senegal, are exposed to high evapotranspiration rates at the end of the rainy season. This results in crust formation. It is necessary to break this crust before the rainy season in order to prepare a seedbed (Charreau and Nicou, 1971). The desirable aggregate size for most seedbeds is 3 to 12 mm (FAO, 1974), and it is therefore necessary to pulverize aggregates formed from crusted soils. However, susceptibility to erosion is correlated with the fineness of the seedbed, so that a compromose must be achieved between the necessity to break the crust and the need for protection against erosion. Charreau and Nicou (1971) believe that it is climatic aggressivity more than soil erodibility that is responsible for erosion and soil degradation. But it seems that overintensive pulverization by tillage is also a cause of the increase in susceptibility to erosion, regardless of the climatic conditions. Lal (1974) reported that it is the detachability of fine particles by splash because of the predominance of coarse sands in many tropical soils that results in erosion. In the view of the author of this chapter, it is the excessive crumbling of the soil at the beginning of the cultural cycle, when the soil is bare, that is the principal cause of erosion resulting from mechanization. In traditional agriculture, where the erosion is not serious, the soil is tilled only when the seeds are placed into the soil. Perhaps it is desirable, at least theoretically, to consider a fine preparation of the soil only along the seed row.

With soils of higher clay content, the results are not so pronounced. Soils with more than 20 per cent clay usually show relatively slight erosion. Charreau and Nicou (1971) consider that clay soils maintain their structure, while sandy soils tend to compact. That is why sandy soils have to be ploughed to improve root development and establishment and minimize the adverse effects of erosion. Consequently, soils of different texture should be considered separately when comparing the diverse advantages and disadvantages of methods of soil preparation.

Erosion is also affected by the soil moisture content at the time of tillage. More data relating to these effects are needed in tropical humid Africa, and this is especially true as regards the consequences of ploughing.

Furrowing and ridging have been practised by many researchers. Under proper conditions, contour ridges offer a mechanical drag for runoff, and when combined with tie-ridging, effectively increase the time for rainwater to infiltrate. However, many observations indicate that in addition to rapid drying, degradation of the upper part of the ridge occurs under the effect of splash, resulting in separation of sand and clay particles and deposition of sand. After some years, the upper horizon is significantly modified by erosion of the surface layer. This process results in a progressive destruction of the ridges due to structural degradation and, in spite of the reduced runoff, the consequences of its use on the structural stability of the soil should be carefully considered. Once more, it is in the sandy and sandy-loam soils that structural degradation is more pronounced.

Whatever the techniques used, mechanization, with or without local modification of the micro-relief, results in a drastic change in the aggregate size distribution. While it may be better to decrease aggregate size to facilitate germination and seedling establishment, one must avoid overpulverization to keep soil erosion under control. The choice of techniques to be adopted, therefore, depends on the soil texture and the moisture content at the time of tillage. But there is another important factor: the depth of tillage. In several experiments, tillage to constant depth has resulted in the formation of a plough-pan, especially in the '*sols ferrallitiques*', where the clay content increases progressively from the A to B horizon. Under some conditions this plough-pan can increase erosion and compaction, resulting in temporary waterlogging and poor soil aeration.

All these disadvantages can be eliminated or minimized by altering the plough depth every year, by avoiding overpulverization of soil, by limiting the tillage to the planting row, and by performing tillage operations only under optimum soil moisture conditions. But these are theoretical conclusions, because in practice it is always difficult to choose the date of ploughing as a function of soil–water content. However, choice of plough-depth in terms of the stage of the cultural succession is usually possible.

Secondary tillage involves inter-row cultivation during the crop growth. These tillage operations are designed to break the upper crust, eliminate weeds and increase soil permeability. Such inter-row cultivation must be done on the contour, to avoid concentrated runoff which can result in serious soil erosion.

A last group of tillage operations concerns post-cultural or post-harvest techniques. Some of these practices such as mulching are very efficient in preventing soil loss. Practices such as ploughing at the end of the cycle, as Charreau has shown, have the advantage of improving infiltration at the start of the rains, when the erosive storms fall on bare but not crusted soil. Post-harvest or early tillage may also help in saving time during the periods of peak labour demand at planting. In regions of short rainy periods, timely planting facilitated by post-harvest tillage can result in significant yield increases, because this technique enables rapid crop establishment and vigorous growth. Some biological techniques are efficient in erosion control (Chapter 4.1). It is desirable to develop mechanized techniques which aid favourable biological processes, such as those which accelerate the establishment of a vegetal cover. That is why it seems necessary to recommend more than one method, such as mulching, strip-cropping and ploughing at the end of the cycle, along with balanced fertilizer use.

5.5.3. CONCLUSIONS

In certain conditions, intensive mechanization of tropical agriculture presents a real danger of accelerated soil erosion.

(i) Mechanical clearing of the soils modifies the micro-climate and decrease resistance to erosion. Adverse effects can be minimized by appropriate methods of root ploughing and by creation of a homogeneous cultural profile, by the use of leguminous cover crops after forest removal.

(ii) Soil preparation techniques must be chosen taking into consideration soil texture, and when ploughing is adopted, the effects if it is done at the beginning or end of the rainy season. One should avoid overpulverization of soil and change the depth of ploughing every season. Investigations of the effects of soil moisture at the time of ploughing on structural stability need to be carried out on soils of different textures.

(iii) Whenever feasible, ground cover should be maintained by practices such as mulching. New research is also necessary to define the consequences of diverse tillage techniques on the structural stability of the soil, on the modification of the porosity of soil aggregates (quantitative and qualitative), and on the development of a cultivated horizon of favourable tilth, merging with the adjacent lower horizon.

(iv) Mechanization presupposes the use of other anti-erosive techniques of contours and terraces, strip-cropping, mulching, and restriction of tillage to the seeding row.

REFERENCES

Bertrand, R, 1967, Etude de l'érosion hydrique et de la conservation des eaux et du sol en pays Baoulé, *Colloque sur la Fertilité des Sols Tropicaux, Tananarive*, 19-25/11/67, No. 106, pp. 1281–1295.

Charreau, C, 1969, Influences des techniques culturales sur le développement du ruissellement et de l'érosion en Casamance, *Agron. Trop.*, **24**, 836–42.

Charreau, C., and R. Nicou, 1971, L'amélioration du profil cultural dans les sols sableux et sablo-argileux de la zone tropicale sèche Ouest-africaine et ses incidences agronomiques, *Agron. Trop.*, **26**, 903–78.

Charreau, C., and L. Seguy, 1969, Mésure de l'érosion et du ruissellement à Séfa en 1968, *Agron. Trop.*, **24**, 1055–1097.

Cointepas, J. P., 1958, Bilan des études chimiques et pédologiques entreprises à la Station expérimentale de Séfa, ORSTOM, Dakar.

Cunningham, R. K., 1963, The effect of clearing a tropical forest soil, *J. Soil Sci.*, **14**, 334–335.

Datiri, B. T., 1974, Soil and water conservation practices particularly in areas of shifting cultivation, *F.A.O. Soils Bulletin*, No. 24, 237–241.

Dumas, J., 1965, Relation entre l'érodibilité des sols et leurs caractéristiques analytiques, Cah. ORSTOM, sér. Pédol., **III**, No. 4, 307–333.

FAO, 1974, Soils Bulletin No. 23, *Management Properties of Ferralsols.*

Fauck, R., C. Moureaux and C. Thomann, 1969, Bilans de l'évolution des sols de Séfa (Casamance, Sénégal) après quinze années de culture continue, *Agron. Trop.*, **24**, 263–301.

Goujon *et al.*, 1968, Conservation des sols en Afrique et à Madagascar. III. Influence des rotations et des pratiques culturales sur le ruissellement et les pertes en terres à Madagascar, *Bois-Forêts Trop.*, **120**, 3–14.

Henin, S., R. Gras and G. Monnier, 1969, *Le Profil Cultural*, 2nd edn, Masson.

Lal, R., 1974, Soil erosion and shifting cultivation, *FAO Soils Bulletin*, No. 24, 48–71.

Le Buanec, B., 1971, Dix ans de culture motorisée sur un bassin versant du centre Côte d'Ivoire—Bilan général, Multigr. IRAT/CI, Bouaké.

Pereira, H. C., P. Hosegood and M. Dagg, 1967, Effects of tied ridges terraces and grass leys on a lateritic soil in Kenya, *Exp. Agric.*, **3**, 89–98.

Roose, E., 1973, Dix sept années de mesures expérimentales de l'érosion et du ruissellement sur un sol ferrallitique sableux de Basse Côte d'Ivoire, Contribution à l'étude de l'érosion hydrique en milieu intertropical, Thèse, Faculté des Sciences de l'Université d'Abidjan.

Russell, E. W., 1971, Soil structure: its maintenance and improvement, *J. Soil Sci.*, **22**, 137–151.

Seguy, L., 1969, Mésure de l'érosion et du ruissellement à Séfa en 1968, *Agron. Trop.*, **24**, 1055–1097.

See also:

Charreau, C. and R. Fauck, 1970, Agricultural development of soils of the Sefa region, *Agron. Trop.*, **25**, 151–191.

Christoi, R., 1966, Erosion measurements in Upper Volta, *Oléagineux*, **21**, 531–534.

IFAC, 1955, Sloping terraces for interception and removal of water, *Fruits*, No. 7.

5.6

Studies of Cultivation Techniques at Bouake, Ivory Coast

J. M. KALMS

5.6.1. INTRODUCTION

The introduction of mechanization is one of the bases of the intensification of the production of arable crops in the Ivory Coast. These large-scale agricultural schemes have been initiated in the last five years, more so in the central region of the country. Although growing arable crops on alternate contours with 5-m wide grass strips in between is adopted as an erosion control practice, a lot of damage to soil and crop is still observed on these farms. It seems that fighting erosion with this system is a losing battle, since erosion is serious even on gentle slopes (5 per cent or less). It is therefore necessary to understand the effects of soil properties and management techniques on runoff and erosion. The influence of different management techniques on runoff and soil loss has been investigated at the research station of Bouake from 1971–1974. The significant results of these studies will be presented in this report.

5.6.2 GEOGRAPHIC LOCATION AND ECOLOGICAL ENVIRONMENT

The research station is located in the central region of Ivory Coast (5° 4′ W and 7° 44′ N). The vegetation is of guinea savannah or secondary forest type, with climate of Baouleen-Dahomeen type. The growing season is characterized by two rainy seasons, separated by a short dry spell in July–August. Rains start mid-March and end early November. The mean annual precipitation is 1192 mm, with a peak in September (212 mm). Evapotranspiration is more than precipitation for five months, annual ETP (Turc) being approximately 1480 mm.

Soils are derived from granitic parent material and represent about 65 per cent of the soils of the central region. The landscape is rolling, with the majority of slopes ranging between 2 and 7 per cent. The surface soil has a coarse to medium texture (10–20 per cent clay) and a frequently high gravel content (10–20 per cent). Gravel content increases with depth up to 50–80 cm and can be higher than 50 per cent. The lower members of the catena are gravelly sands.

Soil organic matter content is generally good, a mean value of 2·2 per cent under cultivation. The moisture-storage capacity of these soils is low to medium (50–80 mm per 100 cm). Though permeability is generally high, crust formation on the soils increases runoff losses.

Runoff plots were established on '*sol ferrallitique*'. The skeletal material in the surface horizon is generally more than 20 per cent. The subsoil is very gravelly (70 per cent) and of sandy-clay texture up to 60 cm below the surface. This soil series is the most dominant soil of the region. Soil erodibility investigations have been conducted on the same location since 1960. The *K* factor is approximately 0·05, which is probably significantly influenced by gravel content and therefore low.

5.6.3. LAYOUT OF THE EXPERIMENT

There are four plots of 112 m^2, 22·5 m long and established on 4 per cent slope. Plots are bordered by asbestos edging. The runoff water passes over a flume into two interconnected storage tanks of 2 m^2 volume each. Four separate treatments are allocated in each plot. A daily-recording raingauge is installed in the vicinity.

5.6.4. EXPERIMENTAL DESIGN

The purpose of the experiment was to investigate the influence of the following treatments:

(i) ploughing up and down the slope to a depth of 25–30 cm (P_1);
(ii) ploughing followed by rotovation (P_2);
(iii) rotovation up to 5–10 cm depth (P_3);
(iv) no tillage (P_4).

All plots were kept free of weeds with herbicides. Tillage treatments were imposed twice a year, in April and August, at the times of optimum soil moisture.

The quantity of runoff in each tank was monitored after every rain. Runoff samples for sediment density were taken only if runoff was more than 5 mm.

5.6.5. RAINFALL CONDITIONS

The Wischmeier (Masson, 1971) erosivity index (*R*) is computed every year from the rainfall records. The annual precipitation and *R* values are shown in Table 5.14.

The rainfall in 1972 and 1973 was below average, 1971 was above average and 1974 was just about an average year. There was a good correlation between rainfall amount and erosivity index (*R*). The mean annual value of *R* from 1960 to 1974 is about 484. Though this value is high, it is smaller than that of the Western region (840) and coastal zone (1260) at Abidjan (Roose, 1973).

Table 5.14. Rainfall and erosivity

Year	1971	1972	1973	1974
Rainfall (mm)	1345	965	959	1121
R factor (US units)	523	329	352	464

5.6.6. PRINCIPAL RESULTS

The measurements in no-tillage plots were initiated only in August 1971.

5.6.6.1. Comparison of Runoff

Table 5.15 gives the mean annual runoff values.

Table 5.15. Mean annual runoff (in % of the erosive rainfall amount)

Plots	P_1	P_2	P_3	P_4
1971	34	32	35	—
1972	37	34	37	42
1973	35	40	47	49
1974	31	31	36	45

In 1971 and 1972, there were no differences between the three tillage plots. Runoff was always higher on the no-tillage plot, and the number of erosive rains for each plot is different for 1973 and 1974 (Table 5.16).

Table 5.16. Number of erosive rains per plot

Plots	P_1	P_2	P_3	P_4
1971	7	8	7	(3)
1972	11	10	11	11
1973	15	12	19	25
1974	13	13	21	26

Taking into consideration the number of erosive rains on the no-tillage plot P_4 for the calculation of the runoff for the other plots, lower runoff is obtained for 1973 and 1974 (Table 5.17).

Table 5.17. Mean annual runoff, in % of the erosive rainfall amount for P_4

Plots	P_1	P_2	P_3	P_4
1973	27	28	42	49
1974	20	21	36	45

Ploughing increased infiltration, reducing runoff significantly. However, breaking crust by ploughing reduced runoff only slightly in comparison with the no-tillage plot.

It is interesting to compare the mean runoff for the first three months after the tillage with the runoff from rains after the first three months and that until the next tillage (Table 5.18).

Table 5.18. Comparison of runoff before and after the soil tillage

Year	Plots	P_1	P_2	P_3	P_4	*R* factor (US)
1973	Before tillage	24	27	46	51	113
	After tillage	27	27	38	47	195
1974	Before tillage	26	27	38	43	215
	After tillage	18	19	35	47	208

The effect of soil tillage was high in 1974 when, for the same *R* before and after the soil tillage, the runoff was less during the three months following the tillage operations.

In 1973, runoff was identical for the two periods but the *R* factor was higher after soil tillage.

Ploughing has a more favourable effect during the first three months of the tillage operation.

5.6.6.2. Comparison of Soil Losses

Table 5.19 gives the total annual soil losses per plot.

Table 5.19. Soil losses (tonnes/ha)

Plots	P_1	P_2	P_3	P_4
1971	11·5	14·9	12·9	—
1972	19·7	11·0	25·0	17·9
1973	17·6	9·3	48·6	41·1
1974	12·2	11·2	43·8	51·9

Ploughing followed by harrowing significantly decreased soil loss in comparison with no-tillage (P_4) or superficial tillage (P_3) treatment. It is interesting to note that soil loss increased with time for P_3 treatments. Harrowing therefore has only a temporary effect and had as much soil loss as no-tillage.

Ploughing followed by harrowing seems to be the best of the cultural techniques compared in this investigation. It should be kept in mind that the no-tillage treatment in this study is without crop residue or mulch. No-tillage with mulch is an important anti-erosive technique (see Chapter 4.3).

5.6.6.3. Comparison of Sediment Density

Turbidity measurements were made for each erosive rainfall. Table 5.20 shows the differences amongst treatments.

Table 5.20. Mean annual turbidity (g/l)

Plots	P_1	P_2	P_3	P_4
1973	0·69	0·73	0·68	0·21
1974	0·58	0·34	0·57	0·26

Turbidity is least for the no-tillage treatment; sometimes one-third or one-quarter of this ploughed treatment may be explained by the existence of a crust and a high proportion of coarse sands on the soil surface of the 'no-tillage' plots. In contrast, the same protecting effect cannot be seen in the 'tillage' plots.

5.6.7. DISCUSSION: MODIFICATION OF THE *K* FACTOR OF THE SOIL

5.6.7.1. Comparison of the Coarse Particle Content of the Upper Soil

Changes in the coarse fraction of the upper 5 cm of the soil have been followed for two years (1973 and 1974). Table 5.21 gives the percentage of coarse fraction in each plot.

Table 5.21. Percentage coarse particles in the upper 5 cm of the soil

Plots	P_1	P_2	P_3	P_4
%	28	22	13	10

Ploughing brought gravels from the subsoil to the surface, while the gravel content for the 'no-tillage' plot was lower. Ploughing followed by harrowing had slightly less gravel content than ploughing alone.

The gravel content of the surface soil is an important factor influencing erodibility of the soil. These results obtained in the last two years are significantly influenced by gravel concentration.

Using the tables of J. Dumas (1965) to determine the *K* factor of the soil for the different plots, we obtain the following values (Table 5.22).

Table 5.22

Plots	P_1 and P_2	P_3	P_4
K	0·05–0·08	0·15–0·18	0·18–0·22

The K value for the ploughed plots corresponds very well to that measured for the soil during the preceding years.

On the plots P_3 and P_4, the decrease of the gravel content of the soil surface seems to be due to the formation of a crust. This evolution of the 'no-tillage' soil could be seen up to the third experimental year.

5.6.8. CONCLUSION

Ploughing results in a considerable decrease in soil loss and runoff in gravelly ferrallitic soils. Its effect increases when it is followed by a slight harrowing.

Superficial pulverization is not desirable.

A basic 'no-tillage' soil without crop residue mulch becomes rapidly susceptible to erosion by the formation of a crust and a decrease in gravels. Gravel content in the surface plays an important role in erosion control.

The introduction of ploughing as a cultural intensification tool in the central region of the Ivory Coast seems to be favourable in maintaining the potential of the ferrallitic soil. However, there is an urgent need to investigate the applicability of mulching and no-tillage techniques with mulch to runoff and erosion. If done properly, no-tillage with mulch may be even better.

REFERENCES

Bertrand, R., 1967, Etude de l'érosion hydrique et de la conservation des eaux et du sol en pays baoulé (Côte d'Ivoire), *Colloque sur la Fertilité des Sols Tropicaux, Tananarive, 1967*, pp. 1281–1295.

Dumas, J., 1965, Relation entre l'érodibilité des sols et leurs caractéristiques analytiques, Cahiers ORSTOM, Pédologie, **III**, No. 4.

Masson, J. M., 1971, L'érosion des sols en climat méditerranéen. Méthodes expérimentales pour l'étude des quantités érodées à l'échelle du champ, Thèse, Université de Montpellier.

Roose, E., 1973, Dix sept années de mesures expérimentales de l'érosion et du ruissellement sur un sol ferrallitique sableux de basse Côte d'Ivoire, Thèse, Université d'Abidjan.

PART 6

Farming Systems and Erosion Hazard in the Far East

6.1

A Brief Review of Erosion Research in the Humid Tropics of South East Asia

R. LAL

The problem of soil erosion in the humid region of South East Asia has to be related to land forms, vegetal cover and methods of land cultivation, viz., shifting cultivation, plantation crops, mechanized agriculture, paddy rice water control system, terracing (Plate 27). Soil erosion problems under shifting cultivation, where adequate time is allowed for fertility rejuvenation, are not serious. The soils for paddy cultivation, the most important crop for South East Asia, are also free from the erosion hazard because of their flat topography and location in the toposequence. The soil erosion problem is accelerated when native vegetation is removed in order to introduce large-scale plantations for rubber, coffee and other tree crops. Soil erosion has been serious in the initial stages of establishment of plantation crops, when forest removal for mechanized farming brought widespread erosion problems (Table 6.1). The Governments of some countries in the region have enacted legislation to enforce the use of soil conservation practices.

6.1.1. SRI LANKA

6.1.1.1. Soil Erosion and Plantation Crops

The landscape of the central highlands of Sri Lanka has been completely changed since the introduction of tea and coffee plantations in the early nineteenth century. Tea plantations are established up to an altitude of 2000 m. Natural vegetation of approximately 200,000 ha was destroyed to establish plantation crops (Manshard, 1968). Clearing of the native vegetation under high-intensity tropical rainstorms caused serious erosion. The earlier reports on soil erosion have been documented by various workers (Lester-Smith, 1939; Burns, 1947; de Rosayro, 1947). Holland and Joachim (1933) stressed the use of cover crops in these plantations (Table 6.2). *Clitoria cajanifolia* and *Indigofera endocephylla* reduced the soil loss from tea plantations as compared with unprotected controls. Burns (1947) estimated that about 25 million tonnes of soil is being washed off the surface of Sri Lanka each year into the sea. He therefore recommended not only that the plantation areas should be guarded against the runon, but also that the

Table 6.1. Suspended sediment loads of some South East Asia Rivers (after Douglas, 1968)

River	Location	Drainage area (km^2)	Suspended load ($m^3/km^2/yr$)
Tjatjaban	Java	79	2500
Tjiloetoeng	Java	620	1352
Dry-zone rivers	Burma	92,500	507·0
Chindwin	Burma	114,500	347·0
Irrawaddy	Prome, Burma	367,000	309·1
Rambut	Java	45	200·0
Mekong	Vientiane, Laos	299,000	164·0
Mekong	Stung Treng, Cambodia	534,000	130·0
Bertam	Cameron Highlands, Malaysia	73	103·1
Irrawaddy	Mandalay, Burma	160,000	75·5
Gombak	Kuala Lumpur, Malaysia	140	67·3
Selangor	Battambang, Cambodia	3230	23·3
Telom	Cameron Highlands, Malaysia	77	21·1
Kompong Lar	Thnot Chum, Cambodia	420	19·8
Sen	Kompong Thom, Cambodia	13,670	16·5
Pursat	Pursat, Cambodia	4480	14·8
Babaur	Babaur, Cambodia	870	11·3
Staung	Kompong Chen, Cambodia	1895	10·9
Krakor	Krakor, Cambodia	138	8·0

runoff water from the plantation crops should be disposed of in the safest possible manner. Maintenance of both natural and artificial drainage systems is therefore important.

Table 6.2. Soil erosion under leguminous cover crops during a five-year period in tea grown on a slope of 30° at Perideniya, Sri Lanka (Holland and Joachim, 1933)

Treatment	Total soil loss (tonnes/ha)	Thickness of soil layer lost (mm)
Control	198·4	21
Clitoria sp. (hedges)	170·2	19
Indigofera (creeper)	101·9	11

6.1.1.2. Research on Soil and Water Loss and Rainfall Patterns

Quantitative studies on runoff and soil loss in food crop production were initiated by Alles (1958). He reported the importance of cover crop in the rotation on soil and water conservation. He concluded that the high-intensity rains of the *Maha* season cause soil detachment and surface sealing on clean-cultivated soil; reduced infiltration and increased runoff and soil loss follow.

Even a mulch rate of only 1·5 tonnes/ha significantly reduces runoff and soil loss. Alles (1971) also tried soil conditioners for erosion control. Curasol (PVA) treated plots had a significantly lower runoff and soil loss (Table 6.3). He proposed that although cover crops and mulching are two of the most effective and widely used methods for controlling soil erosion, there may be times when chemical sprays which bind the soil aggregates would be helpful either as a supplement to cover crops and mulching or instead of them.

Table 6.3. Role of soil conditioners in erosion control (Alles, 1971)

Treatment	Soil splash (g/sample) From 6 rainstorms between Feb. 25 & Mar. 29	Soil splash (g/sample) From 6 storms between Mar. 30 & Apr. 25
Curasol-treated	19·0	24·0
Control	40·3	31·7

Quantitative analyses of soil erosion records under tea have also been reported by Hasselo and Sikurajapathy (1965). Bare plots under tea suffered the greatest soil loss, while the tea planted after Guatamala grass had tolerable loss. Bare plots lost 23·5 tonnes/acre of soil over two and a half months. Mulched plots lost 10–9 tonnes/acre of soil over the same period.

These workers calculated the erodibility (K) of tea soils for different storms (Table 6.4), using the Wischmeier equation. In general, the erodibility was low when the tea canopy was almost full. It is obvious from the data that soil erodibility K is both a function of soil characteristics as well as management practices. There is a 30 times reduction in K by increasing vegetal cover.

Table 6.4. Soil erodibility (K) values (after Hasselo and Sikurajapathy, 1965)

Treatment	Date	Rainfall (cm)	Soil loss (tonnes/ha)	R (ft-ton/ acre-in.)	LS	K
Bare	28/4	3·30	14·42	9·95	4·6	0·31
Bare	13/5	4·57	3·10	10·40	4·6	0·31
Tea	28/4	3·30	2·10	9·95	3·8	0·05
Tea	13/5	4·57	0·41	10·40	3·8	0·01

The rainfall patterns, distribution and intensity records of Sri Lanka for different regions have been published by Alles (1958, 1971) and De Mel (1971). Sri Lanka has been divided into four regions according to the rainfall distribution (Sirimane, 1958). The regional analysis and the runoff from each of the regions are given in Table 6.5. The mean annual rainfall of the island is 202·2 cm. It is estimated that 25 per cent of it is lost as runoff.

Table 6.5. Land classification and percent runoff (Sirimane, 1958)

Zone	Rainfall (cm)	Runoff (%)
Wet zone	250–500	71–84
Humid zone	190–250	47–67
Dry zone	125–190	25–47
Arid zone	50–125	0–12

6.1.1.3. Research on Mulch-tillage Techniques

The use of mulch for erosion control has been well recognized in this island. Livera (1947) reported that use of crop residues as surface mulch is helpful in intercepting and dispersing raindrops, preventing the clogging of pores and surface compaction or crusting of the surface soil. Livera reported that surface runoff and evaporation losses are less with mulch than when crop residues are incorporated in the soil. The importance of crop residues in soil and water conservation was further strengthened by the research conducted at the 'dry zone' research station at Maha Illupallama (Abeyratne, 1956). Abeyratne reported that flat cultivation with trash cover improved the soil moisture regime and increased the yield of chillies and cotton.

Ridge cultivation was generally a poor yielder. Trash mulches increased the chillies yield by 24 per cent and cotton yield by 38 per cent as compared to flat-cultivated unmulched plots.

The general problem with the mulch-farming techniques was weed control (Kirinde, 1957). If the weeds could be successfully controlled, many cultivation operations involving overexposure of the soil could be avoided. Kirinde reported that unweeded plots with no cultivation had as much as 55 per cent yield reduction in upland rice as compared with cultivated plots with better weed control (Table 6.6).

Table 6.6. Yield reduction in upland rice by weeds (Kirinde, 1957)

Treatment	Yield (kg/ha)	% of control
Pre-sowing treatment		
No cultivation at sowing time	1301	100
Cultivated at sowing time	2021	155
Time of first weeding		
10 days after sowing	1944	100
15 days after sowing	1666	84
20 days after sowing	1325	66
Frequency of weeding		
3 weedings at 10-day interval	1614	100
6 weedings at 10-day interval	1708	106

Kirinde's experiments indicated that the use of mulches with herbicides holds the greatest promise of weed control and soil and water conservation in the rainfed areas of Sri Lanka.

6.1.2. MALAYSIA

6.1.2.1. Soil and Water Loss

The erosion problem originates from changes in the land-use pattern which have drastically affected the hydrological cycle and the physical, chemical and biological properties of the soil. The erosion will continue to increase as new land is brought under agriculture by forest clearing. Out of the total area available for development, 1 million acres of forest land were to be cleared and developed for agriculture during 1971–75.

Hydrological investigations in Malaya giving the details of the rainfall and river discharge from four major rivers have been reported by Oh Kong Yew (1965). Forest removal will significantly alter the hydrological balance and the sediment-carrying capacity of the discharge.

Soil losses from catchments under various vegetation covers differed. In the highlands with hilly catchments, the erosion losses were:

Forest	245 m^3/km^2 annum
Tea	488 m^3/km^2 annum
Vegetables	732 m^3/km^2 annum

If 1 m^3/km^2 annum is approximately a loss of 0·01 mm of soil, vegetable production will result in an equivalent of 7·32 mm of soil per year. Similar estimates under rubber and oil palm are given below:

Undisturbed jungle catchment	31 m^3/km^2 annum
Mature rubber and oil palm (small catchment)	46 m^3/km^2 annum
Mature rubber and oil palm (large catchment)	87 m^3/km^2 annum

Though there will be serious soil erosion during plantation establishment, tree crops are preferred over arable crops if erosion is a serious factor.

From the point of view of land development and erosion control in the plantation, the role of cover crops has long been recognized by Malayan agricultural authorities. Bunting and Milsum (1928) recommended the use of various cover crops on rubber, coconut and oil palm estates throughout Malaya. In addition to controlling erosion on hilly and undulating slopes, these cover crops also reduce weeding costs, decrease soil temperature, and maintain soil structure, organic matter content and the fertility of the soil. However, they also recognized various objection to these cover crops. For instance, (a) twining plants damage the permanent crops, (b) there is a risk of fire, (c) there is a danger of harbouring diseases and insects. Some of the recommended cover crops included *Calopogonium mucunoides*, *Centrosema pubescens*, *Dolichos hosei*, *Indigofera endocephylla* and *Pueraria phaseoloides*.

The effect of various cover crops on the soil moisture regime showed that they neither increased nor significantly diminished the soil moisture regime under weather conditions ranging from moderately wet to very dry. Nevertheless, there were considerable reductions in water loss from the surface runoff. The results of the National Water Resources Survey of Malaysia have been given in a report by Lim (1974).

6.1.3. PHILIPPINES

Erosion research in the Philippines has not been extensively reported. Kellman (1969) reported runoff and soil loss studies from the upland Mindanao region under shifting cultivation. The average percent runoff under different vegetation is shown in Table 6.7. The lowest runoff was recorded beneath primary forest and

Table 6.7. Runoff from plots with different vegetation (Kellman, 1969)

	Percent runoff during:		
		Uncropped period	
	Cropping period	Pre-cropping	Post-cropping
Primary forest	0·26	—	—
Softwood tree fallow	0·26	—	—
Imperata grassland	$3{\cdot}0^{+}$	—	—
New abaca plantation	0·35	—	—
10-yr-old abaca plantation	0·64	—	—
Logged-over forest	—	1·73	—
New corn swidden	1·52	—	0·86
New rice swidden	1·08	0·69	0·42
Two-yr-old corn swidden	1·78	—	4·08
12-yr-old rice swidden	11·64	6·73	14·15

Table 6.8. Effect of vegetation cover on soil loss (after Kellman, 1969)

	Soil loss during:	
Vegetation cover	Cropping period (g/day)	Plant harvesting period (g/day)
Primary forest	0·20	—
Softwood tree fallow	0·29	—
Imperata grassland	0·40	—
New abaca plantation	0·47	—
10-yr-old abaca plantation	0·59	—
New corn swidden	3·03	0·65
New rice swidden	1·45	0·37
Two-yr-old corn swidden	12·05	9·81
12-yr-old rice swidden	119·31	6·32

softwood tree stand. High runoff under grassland is related to leaf arrangement permitting ready raindrop penetration, indicating the deleterious effects of changing forest vegetation to grassland savannah. Runoff from corn was more than from rice.

The lengthening of rice cultivation to a 12-year period significantly increased runoff from 1·08 per cent to 11·6 per cent, which is an index of deterioration of soil structure under continuous cultivation. The soil loss under different vegetation cover is shown in Table 6.8. The plot area was 8 m^2. Soil erosion increased as much as 100 times from new and old rice plots.

6.1.4. THAILAND

The total area of the mainland is 500,000 km^2. The more serious erosion problems are in the northern highland region because of the rainfall pattern and land forms. Tin mines have also spoiled about 10,000–30,000 ha of land surface. The area is currently cultivated under the system of shifting cultivation (Anthony and Moormann, 1965). It is estimated that 16 to 32 thousand ha of land in the northern provinces is cleared annually for cultivation. The total watershed area under shifting cultivation is 114,000 km^2. While there is no serious erosion problem under low intensities of land use by shifting cultivation, erosion becomes serious when cultivation is prolonged or large-scale forest-clearing schemes are introduced without adequate attention to erosion control.

Quantitative studies on the magnitude of soil erosion in Thailand are few. Anunt Komes (1972) reported that forest clearing on sandy soils led to a change in vegetation from forest to savannah, with *Imperata cylindrica* as the dominant grass and severely eroded slope lands. The annual runoff losses were 10 mm and 20 mm under forest and *Imperata cylindrica*, respectively. The annual soil loss varied from 0·064 tonnes/ha to 0·075 tonnes/ha under the uncultivated forest or savannah vegetation. Surface runoff and sediment loss from natural forest is also reported by Tangtham *et al.* (1972). Agronomic research has indicated that the use of herbicides has also shown encouraging results.

6.1.5. OTHER ASIAN COUNTRIES

Only part of southern India comes within the humid tropical region. References to erosion studies in India are given at the end of this chapter, but most of these refer to the semi-arid zones.

Little erosion research in Indonesia has been reported, although the rainfall intensity records of various locations in Indonesia indicate the potential erosion hazard to be high (Table 6.9). Lester-Smith (1939) reported that in Java, for altitudes of over 1000 m and receiving more than 3200 mm of rain in the year, 60 per cent or more of the rainfall was lost as surface runoff. The effect of soil conditioners on the soil stability of ash emitted from volcanic eruptions has been studied by Muljadi *et al.* (1972). With a simulated rainfall intensity of 200 mm/hr on a slope of 7 per cent during 90 minutes, the amount of eroded soil from the untreated sample was equivalent to 25 tonnes/ha and 0·25 tonnes/ha for the treated soil. The analysis of soil erodibility of different soils using a rainfall simulator has also been described by Keersebilek *et al.* (1972).

Table 6.9. Maximum rainfall intensity for various locations in Indonesia (after Keersebilek *et al.*, 1972)

Duration (min)	Maximum intensity (mm/hr) Jakarta	Bogor	Pasuruan	Padeng	Pontianak	Ambon	Mean
5	192	192	192	192	228	216	204
15	140	156	140	172	160	156	156
30	118	130	128	152	146	96	128
60	88	110	86	100	80	63	88

Laos is a landlocked mountainous country with 23·6 million hectare of land out of which only 2 m hectare is considered suitable for agricultural use. The country is mountainous, with altitudes as high as 2000 m above sea level. Out of the total rain received of 12 km^3, 8·8 km^3 is lost as surface runoff. This much runoff results in a serious erosion hazard (Choulamountry, priv. comm.).

REFERENCES

Abeyratne, E. L. F., 1956, Dry land farming in Ceylon, *Trop. Agricst., Ceylon*, **113**, 191.

Agarwal, M. C., R. P. Agarwal and R. S. Chaudhry, 1974, Infiltration characteristics of soils as related to soil physical properties, *J. Indian Soc. Soil Sci.*, **22**, 285–289.

Alles, W. S., 1958, Some studies on runoff and infiltration, *Trop. Agricst., Ceylon*, **114**, 197.

Alles, W. S., 1971a, The rainfall data of Ceylon, *Trop. Agricst., Ceylon*, **120**.

Alles, W. S., 1971b, Chemical sprays for controlling soil erosion, *Trop. Agricst., Ceylon*, **127**, 179.

Anthony, K. R. M., and F. R. Moormann, 1965, Agricultural problems and potentialities of a Hill Tribe Area in Thailand, *Trop. Agric.*, **42**, 97–104.

Ayres, Q. C., 1936, *Soil Erosion and its Control*, McGraw-Hill, New York, p. 83.

Bhattacharjee, J. C., 1957, Erosion study in the laterite area of West Bengal, *J. Indian Soc. Soil Sci.*, **5**, 229.

Bishan Mansingh, Erosion and its control, *J. Indian Soc. Soil Sci.*, **5**, 229.

Bunting, B., 1928, Notes on Dolichos Hosei Sym. Vigna Hosei, *Malayan Agric. J.*, **14**, No. 5.

Bunting, B., and T. D. Marsh, 1925, Observations on cover crops at Castleton Estate, *Malayan Agric. J.*, **13**, No. 10.

Bunting, B., and J. N. Milsum, 1927, Observations on the more important cover plants in Malaya, *The Planter*, **8**, No. 2.

Bunting, B., and J. N. Milsum, 1928, Cover crops and green manures, *The Malayan Agric. J.*, **16**, No. 7, 256.

Burns, R. V., 1947, Soil erosion in Ceylon with particular reference to floods, *Trop. Agricst.*, **103**, *Ceylon*, 240.

Digar, S., and A. K. Sen, 1960, Significance of slope and erosion in land capability classification in the Machkund catchment, *J. Indian Soc. Soil Sci.*, **8**, 157.

Douglas, I., 1968, Erosion in the Sungei Gombak Catchment, Selangor, Malaysia, *J. Trop. Geography*, **26**, 1–16.

Govrie, R. M., 1952, Soil conservation notes. Land slides, *Trop. Agricst., Ceylon*, **108**, 127.

Grigg, D., 1970, *The Harshlands*, Macmillan St. Martin, London.

Gupta, R. N., and B. Narain, 1971, Investigations on some physical properties of alluvial soils of U.P. related to conservation and management, *J. Indian Soc. Soil Sci.*, **19**, 11–22.

Hadimani, A. S., and N. G. Perur, 1971, Measurement of infiltration rate in some of the soils of the Tungabhadra Project Area (Mysore State), *J. Indian Soc. Soil Sci.*, **19**, 231–235.

Haridasan, M., and R. K. Chibber, Effect of physical and chemical properties on the erodibility of some soils of the Malwa plateau, *J. Indian Soc. Soil Sci.*, **19**, 293–298.

Harper, R. S., 1971, Efficiency of paraquat as a weed control chemical for young rubber seedlings. I. Safety of paraquat to seedlings prior to budding, *Thai. J. Agr. Sci.*, **4**, 181–185.

Hasselo, H. N., and M. Sikurajapathy, 1965, Estimation of losses and erodibility of tea soils during the replanting period, *J. National Agric. Soc. Ceylon*, **2**, 13.

Holland, T. H., and A. W. R. Joachim, 1933, A soil erosion experiment, *Trop. Agricst. Ceylon*, **80**, 199–207.

Jones, L. A., 1943, Drainage as a conservation practice, *Trop. Agricst., Ceylon*, **99**, 45.

Karunaratne, C. R., 1938, Soil erosion prevention at Nugawela supply nursery and multiplication station, *Trop. Agricst., Ceylon*, **111**, 156.

Keersebilek, N., D. Muljadi and S. Sukmena, 1972, The possible application of a rainfall simulator method for wet tropical conditions, *Proceedings of the second ASEAN Soil Conf.*, Vol. II, pp. 37–52.

Kellman, M. C., 1969, Some environmental components of shifting cultivation in upland Hindano, *J. Trop. Geography*, **28**, 40–56.

Kirinde, S. T. W., 1957, Weed control in dry land farming, *Trop. Agricst., Ceylon*, **113**, 131.

Komes, Anunt, 1972, Reclamation and maintenance of fertility of major soils in Thailand, *Thai J. Agric. Sci.*, **6**, 127–143.

Lester-Smith, W. C., 1938, Soil erosion on tea estates and some suggestions for its control, *Trop. Agricst., Ceylon*, **111**, 280.

Lester-Smith, W. C., 1939a, Soil erosion and soil fertility, *Trop. Agricst., Ceylon*, **112**, 89.

Lester-Smith, W. C., 1939b, Vegetation, climate, and soil and water conservation, *Trop. Agricst., Ceylon*, **92**, 336.

Lim, Cheong Chup., 1974, A national water resource inventory as a guide to agricultural land use potential. Expert consultation on the use of improved technology by farmers in the rainfed areas of Monsoon Asia. Kuala Lumpur 8–13, December, 1974, FAO, Rome.

Livera, E. J., 1947, Agriculture and erosion in Ceylon, *Trop. Agricst., Ceylon*, **103**, 234.

Mabbayad, B. B., and I. A. Buencosa, 1967, Tests on minimum tillage of transplanted rice, *Philippine Agriculturist*, **51**, 541.

Mehta, K. M., V. C. Sharma and P. G. Deo, 1963, Erodibility investigation of soils of Eastern Rajasthan, *J. Indian Soc. Soil Sci.*, **11**, 23.

Mel, I. D. de, 1971, Comparison of rainfall over Ceylon during the two 30 year periods (1911 to 1940) and (1931 to 1960), *Trop. Agricst., Ceylon*, **127**, 1.

Manshard, W., 1968, *Tropical Agriculture*, Longmans, London.

Milsum, J. N., and E. A. Cutler, 1925, *Calopogonium mucunoides*, *Malayan Agric. Journal*, **13**, No. 8.

Muljadi, D., M. De Boodt, R. Vandevelde and J. Lenvain, 1972, Soil conditioning of volcanic debris: results obtained on Java and Bali, *Pedologie*, **22**, 100–122.

Murthy, R. S., S. P. Jain and S. R. NagaBhushona, 1968, Soil survey and land capability classification for sound watershed management in Kundah Project (Modres), *J. Indian Soc. Soil Sci.*, **16**, 223–228.

de Rosayro, R. A., 1947, Forests and erosion—with special reference to Ceylon, *Trop. Agricst., Ceylon*, **103**, 246.

Singh, N. T., and K. S. Verma, 1975, Effect of rainfall intensity and surface condition on water erosion in foot-hill soils of the Punjab, *J. Indian Soc. Soil Sci.*, **23**, 27–30.

Sirimane, C. H. L., 1958, Land classification in Ceylon with reference to basic conditions of soil and water conservation, *Trop. Agricst., Ceylon*, **114**, 29.

Tamhane, R. V., T. D. Biswas, B. Das and G. C. Naskar, 1959, Effect of intensity of rainfall on the soil loss and runoff, *J. Indian Soc. Soil Sci.*, **7**, 231.

Tangtham, N., K. Chunkao and S. Ungkulpakdikul, 1972, Surface runoff and sediment from natural forest and shifting area on mountainous watershed, Research Note No. 10, Faculty of Forestry, Kasetsart University.

Tejwani, K. G., and V. V. Dhruva Narayana, 1961, Soil conservation survey and land use capability planning in the ravine lands of Gujarat, *J. Indian Soc. Soil Sci.*, **9**, 233.

Plate 27. A combination of soil and water conservation measures and appropriate land use in Kerala, South India, provides the basis of productive agricultural development

[*Facing page* 212

6.2

Erosion Hazard and Farming Systems in the Humid Tropics of South Asia

C. R. PANABOKKE

6.2.1. INTRODUCTION AND GENERAL BACKGROUND

The wide range of erosion hazards that are encountered in the humid tropics of South Asia essentially derive from the diversity of rainfall, soils and terrain conditions that obtain within this part of the tropical world.

Rainfall in this geographic area is determined by regional as well as local phenomena. The monsoonal rains occur as a regional phenomenon, while the convectional rains are a local phenomenon. The mean annual rainfall ranges from 5000 mm in the very wet regions to 1250 mm in the semi-humid regions, and it follows a bimodal pattern. The inter-monsoon convectional rains consist of short duration, high-intensity thunderstorms that could, infrequently, attain intensities of up to 100 mm per hour.

Nine out of the ten soil orders (7th approximation) are encountered within this geographic area. The more important and widespread soil orders, however, are the Alfisols, Ultisols, and Oxisols. Terrain conditions range from flat to gently undulating in the Oxisols; undulating to rolling in the Alfisols; and undulating to hilly in the Ultisols.

The erosion problems of the Alfisols have been studied in greater depth than any of the other soil orders in Sri Lanka because these are the lands that constitute our 'new frontiers'. Moreover, in most parts of South Asia, erosion hazards and management problems assume a greater order of importance in the Alfisols.

Traditional agricultural systems in the different agro-ecological regions of South Asia have evolved over a long time-scale within the historic period. Intensive settlement has taken place on the various alluvial plains that have an assured supply of water. Irrigated rice is the main agricultural enterprise in the various alluvial plain land systems.

On the other hand, in the various catenary landscape systems that typify a major part of South Asia, irrigated or semi-irrigated rice culture in the hydromorphic soil associations has been closely linked with varying components of rainfed shifting agriculture in the adjacent well-drained, well-irrigated lands.

This system of land use is well adapted to the soil and hydrological conditions of the various catenary landscape systems of South Asia.

Until the advent of commercial plantation agriculture, the non-irrigable lands were primarily used for raising subsistance crops in a bush fallow system. Commercial plantation crops such as tea, rubber, cocoa, etc., now occupy some of this non-irrigable land, especially in the Ultisols, where the rainfall is found to be adequate for these crops. It is in the Alfisols, however, that the non-irrigable rainfed lands still await the development of minimum cost, but economically viable and socially acceptable farming systems.

6.2.2. GEOMORPHIC SETTING, SOILS ENVIRONMENT AND HYDROLOGY

A greater part of South Asia is floored by the crystalline metamorphic basement rocks of the pre-Cambrian. Tectonic uplift followed by subsequent peneplantation has resulted in the formation of two or more peneplain surfaces. The lowest peneplain which occupies a greater proportion of this geographic area could be described as a gently undulating rock-floored plain with scattered monadnocks breaking the relief.

Later tertiary as well as quaternary morphogenesis has resulted in the development of land surfaces and drainage systems that reflect the influence of more recent events. The presence of pediment plains and stone lines are suggestive of fluctuating climatic conditions during the quaternary. A rejuvenation of the drainage system can also be observed in the broad valleys of the lowest peneplain.

The Alfisols in Sri Lanka occur on a typical catenary landscape where the slope of the land surface averages 2 to 3 per cent. The Rhodustalfs and Haplustalfs occupy the well-drained aspects of the gently convex landscape, and the Tropaqualfs occupy the imperfectly and poorly drained aspects of slightly concave and flat landscape. The meteorological data given in Table 6.10 are for

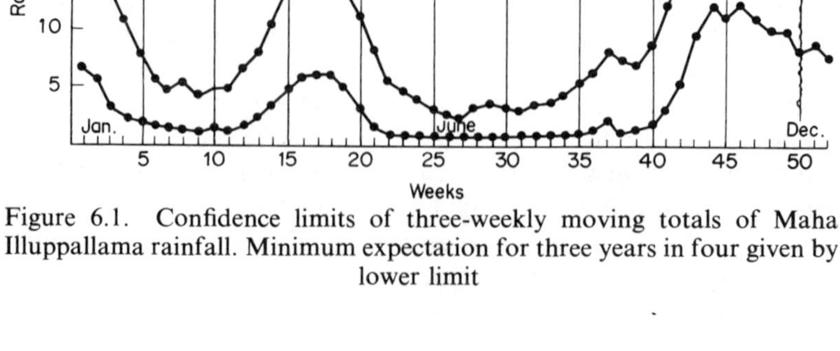

Figure 6.1. Confidence limits of three-weekly moving totals of Maha Illuppallama rainfall. Minimum expectation for three years in four given by lower limit

the Maha Illuppallama Agricultural Research Institute, which is the central research station for the Alfisol region in Sri Lanka. The data given in this table could be taken as indicative of the general environmental characteristics of this region. Rainfall expectation and variability are best expressed in terms of the 1:1 confidence limits of rainfall. Figure 6.1 shows the three weekly moving totals of rainfall confidence limits calculated from a 50-year run of data.

Table 6.10. Meteorological data for Agricultural Research Institute, Maha Illuppallama

Month	Mean rainfall (cm)	Mean evaporation open pan (cm)	Mean temperature Max. (°C)	Min. (°C)	Mean relative humidity	Mean percentage sunshine
Jan.	13·5	13·2	28·2	20·4	81	65
Feb.	4·8	14·2	30·2	20·6	74	75
Mar.	9·9	18·0	32·5	21·8	73	85
Apr.	18·8	17·3	32·8	23·3	76	82
May	9·9	17·8	32·3	24·4	76	77
June	2·8	18·3	31·9	24·4	71	75
July	3·8	19·0	32·3	24·2	70	73
Aug.	4·3	20·3	32·7	23·9	68	78
Sept.	7·6	20·8	32·9	23·6	68	74
Oct.	25·1	15·0	31·3	22·8	77	64
Nov.	26·7	11·7	29·6	21·7	81	56
Dec.	20·3	11·4	28·1	20·9	83	58
Total:	147·6	197·1				

Thunderstorms with intensities of 25 mm in 8 to 15 minutes, and 50 mm in 30 to 40 minutes, have been recorded at the Research Station during the convectional rains that fall during October. The modal value of the minimum infiltration rate after a lapsed time of approximately four hours is 25 mm/hr for the Rhodustalfs. Surface sheet erosion is observed even under a natural forest canopy during the high-intensity thunderstorms that occur immediately following the end of the dry season. Dry weather flow in the second-order streams usually ceases after about 45 cumulative dry days. Very limited quantities of groundwater are available, and that too only where the basement is highly fractured and weathered. Surface water storage in low head reservoirs that are located across the broad valleys of a dendritic drainage system has been the traditional method of water conservation both for domestic requirements as well as for supplementary irrigation.

The Oxisols occur in the same rainfall and climatic environment as the Alfisols. The modal value of minimum infiltration rate is 300 mm/hr for the Oxisols. Even the very high-intensity thunderstorms can be completely accepted by these soils, and no runoff is therefore experienced. As a result, this landscape surface is quite resistant to any form of natural erosion.

The Ultisols occur in the higher rainfall areas and are found both in the lowest peneplain as well as in the highly dissected middle and upper peneplains. The dry seasons are of short duration and the soil moisture profile rarely dries out beyond the 75 cm rooting depth. Dry weather flow generally persists in the second-order streams even after 60 cumulative dry days. The minimum infiltration rate for the Ultisols slightly exceeds the highest rainfall intensity that occurs in this region. Soil erosion is normally absent under a natural forest canopy in the undulating and rolling terrain. Soil creep does, however, occur in the steep and hilly terrain in the very high rainfall regions.

6.2.3. SOIL PHYSICAL PROPERTIES AND EROSION HAZARD

A study of some selected soil physical properties reveals that the erosion hazard would be quite pronounced for the Alfisols, and minimal for the Ultisols. Soil aggregates from the surface horizons of the Alfisols show a very low water stability, while infiltration rates at the steady state rarely exceed 25 mm/hr. Soil aggregates from the Ultisols show a high water stability and they also show very little incipient failure when subject to sudden wetting.

The modal soil member at the Maha Illuppallama Research Station could be described as a Typic Rhodustalf, loamy, kaolinitic, isohyperthermic. The field texture of the surface horizon could be described as sandy clay loam with grit, the coarse sand and grit being mostly angular and subangular quartz. Textural composition is usually 20 to 25 per cent clay; 5 to 8 per cent silt; 25 to 30 per cent fine sand; and 30 to 40 per cent coarse sand. The silicate clay fraction is kaolinitic and the sesquioxide clay fraction is mainly haematite. The microaggregate stability is moderately good, but the structural bonds that bind the microaggregates into the larger microaggregates are very weak and ephemeral. As a consequence, slaking of the microaggregates following sudden wetting is quite rapid. A further special feature of the soil is its soil–water energy relationship, where nearly 85 per cent of the available moisture is lost at a tension of 1 atmosphere or pF 3. As a result, the soils dry out very easily and also tend to get saturated very quickly after rainfall. Surface sealing following the impact of raindrops on a bare exposed soil surface can be unusually rapid on a soil with the foregoing physical properties.

6.2.4. FARMING SYSTEMS AND EROSION HAZARD

In the various agro-ecological regions that are identifiable within the Ultisols, the erosion hazard is minimal. In the well-managed plantation sector, especially tea and rubber, and also in the traditional mixed homegarden with its multistoreyed perennial crops, direct raindrop impact on a bare soil is minimized. In the arable crop sector, where proper bench terracing has been resorted to, safe disposal of water at non-erosive velocities is practically feasible. Because of the good dry weather flow in the streams, year-round cropping is possible on some proportion of the arable crop land. The erosion hazard in this soil region is mainly a social and economic phenomenon, and that too in the steep and hilly terrain that has been recently encroached on by marginal subsistence farmers.

In the Oxisols, the settlement density has been largely governed by the availability of underground water. Where adequate supplies of such underground water are available, it is used very frugally. The erosion hazard is minimal on account of the flat landscape and the high infiltration rate of the soil.

It is in the Alfisols that the erosion hazard is most pronounced. One of the main imperatives of modern agricultural research is, therefore, that of evolving stable and economically efficient farming systems for the different agro-ecological regions that are identifiable within the Alfisols. A case study of the experiences in the Rhodustalfs of Sri Lanka would best exemplify this.

The chief objective of the agricultural research project that was initiated in 1950 at the Maha Illuppallama Research Station was that of the development of economic systems of settled arable farming on lands that were hitherto subject to shifting cultivation. The results of this research project have been discussed by Abeyratne (1956, 1962). It has become quite evident in this study that the erosion hazard should be viewed in relation to the total farming system. In this context, the subject is logically discussed within the framework of the three broad categories:

1. Land layout and settlement patterns.
2. Soil management and cultivation practices.
3. Cropping systems.

The more important elements of these are briefly outlined in the following sections.

6.2.4.1. Land Layout and Settlement Patterns

(a) Each minor drainage basin or micro-catchment unit in the catenary landscape has to be treated as the natural unit of management for purposes of efficient water control, and also to ensure a safe disposal of water at non-erosive velocities.

(b) Within each micro-catchment, a system of graded terraces that discharge the storm waters at non-erosive velocities into a straightened, grassed, natural drainage line has been found to be the most effective method of erosion control.

(c) Surface water storage in small reservoirs would help to maximize the use of the total water resource base.

(d) Settlement in clusters around the periphery of these small storage reservoirs would help to minimize erosion hazards.

6.2.4.2. Soil Management and Tillage Practices

(a) Soil management practices have to cater for the safe disposal of surplus water during excessive rain and, at the same time, cater for moisture conservation during the growing season.

(b) Subunits of management within a micro-catchment would therefore have to embrace multiples of individual graded terraces.

(c) Tillage operations should, as far as possible, be reduced to a minimum. If tillage implements have to be used, these should be of a non-inverting type that leaves a maximum protection of stubble on the surface. Continuous protection of the soil in the form of a crop or a mulch is essential.

(d) Combating the prolific weed growth, which is one of the crucial management problems, demands proper timing of weed control or tillage operations, especially on soils that dry out as rapidly as they tend to get saturated.

(e) The very hard consistency of these soils in the dry state, their highly abrasive qualities on tillage implements, and the high bulk densities that impede root development may together necessitate some novel soil ameliorative techniques.

6.2.4.3. Cropping Systems

(a) Different cropping patterns are needed for the individual drainage associates of the soil catena.

(b) Diversification of cropping at the farm level will therefore be inevitable if different cropping patterns are to be accommodated within the individual drainage associates.

(c) In the intensification of agricultural production and the minimization of the erosion hazard, best use will have to be made of the incident rainfall. This would principally involve both a proper choice of sowing time and a proper selection of sowing-to-harvest duration of crops, so that there is a maximum likelihood of the rainfall satisfying the crop-water demand. More detailed aspects of this have been considered by Panabokke (1974).

(d) Cropping systems that cater for minimum tillage operations and permit maximum periods of soil cover will be more desirable in respect of erosion control.

(e) Irrigated or semi-irrigated rice in the lowermost hydromorphic soil associates will have to be closely linked with the rainfed upland in the total farming system within each micro-catchment unit.

ACKNOWLEDGMENTS

The information on soil physical properties and other soil characteristics has been drawn from both published and unpublished data of the Land Use Division, Irrigation Department, Sri Lanka.

SUMMARY

A wide range of erosion hazards are encountered in the humid tropics of South Asia. The erosion hazard is most pronounced in the Alfisols on account of (a) the low water stability of the soil aggregates, (b) easy slaking of the soil macroaggregates following sudden wetting, (c) rainfall intensities often exceeding the infiltration rates of the soils, and (d) the undulating nature of the catenary landscape systems.

The Oxisols occur on landscape systems that are normally resistant to erosion both on account of the high infiltration rates of the soils and the flat nature of the terrain. The erosion hazard is also low in the Ultisols because of the high water stability of the soil aggregates and the moderately high infiltration rates of the soils.

Soil erosion is easily controlled in the well-managed tea and rubber plantation sectors in the Ultisols, except in the steep and hilly terrain. The prevailing farming systems in the Oxisols are such that little or no soil erosion is encountered. Even with intensification of the present farming systems, soil erosion can be easily controlled in the Oxisols.

In the traditional farming systems that prevail in the Alfisols, erosion hazards are moderate when the population pressures are low. With increasing population pressures on the land, the erosion hazards are aggravated. In the transformation of traditional farming systems in the Alfisols, particular attention has to be paid to (a) land layout and settlement, (b) soil management and tillage practices, and (c) cropping systems, in order to minimize the erosion hazard.

REFERENCES

Abeyratne, E., 1956, *Trop. Agric.*, **62**, 191–229.
Abeyratne, E., 1962, *Proc. Ceylon Ass. Advmt. Sci.*, **18** (2), 58–72.
Panabokke, C. R., 1974, *J. Natn. Sci. Coun. Sri Lanka*, **2** (2), 95–113.

6.3

Hydrologic Properties of Soils in Relation to Runoff and Water Control

S. TIENSEEMUANG and P. PONSANA

6.3.1. INTRODUCTION

Runoff occurs in tropical areas when rainfall intensity is greater than the capacity of soil to absorb water, and the land has sufficient slope to allow water movement.

Factors affecting the infiltration capacity are the condition of the soil surface, the soil moisture content, topography, rainfall intensity and duration, and the permeability of the profile.

Research at the Central Region Agricultural Centre has investigated water movement into the soils on the Central Plain of Thailand. While erosion is not a problem on these particular soils, the basic data obtained can assist in the evaluation of soils with regard to erosion hazard and water management. The research results will be discussed with this aim in mind.

6.3.2. SITES AND SOILS

The four soil series referred to in this paper are in the recent and semi-recent river alluvia; these include the Phetburi (Pb), Nakhon Prathom (Np), Chainat (Cn), and Ratchaburi (Rb) series.

Van der Kevie (1965) described these soils briefly as:

Pb = Phetburi series—a sandy loam soil underlain by sandy clay loam, developed on semi-recent alluvium and located on river levees;

Np = Nakhon Prathom series—a clay loam soil underlain by clay, developed on semi-recent alluvium and located on the transitional zone between the river levees and the river basin;

Cn = Chainat series—a silty loam soil developed on recent alluvium in the same topographical position as the Np series;

Rb = Ratchaburi series—a clay soil developed on recent alluvium and located on the higher parts of the river basins.

6.3.3. METHODS OF INVESTIGATION

6.3.3.1. Surface Infiltration

Surface infiltration was measured by means of infiltrometer rings (30 cm i.d.) that were driven 10 cm deep into cultivated soil in the field. Theoretically, infiltration can be described by the two-parameter equation of Philip (1957) as modified by Talsma (1969) for field soils:

$$I = St^{1/2} + K_0 t/2{\cdot}8$$

where I = cumulative infiltration (cm); S = sorptivity (cm min$^{-1/2}$); K_0 = saturated hydraulic conductivity (cm/min); t = time (min).

6.3.3.2. Vertical Hydraulic Conductivity

Vertical saturated hydraulic conductivity K_0 was measured at 20 cm intervals down the profile by taking 30 cm diameter deep cores from a soil pit and measuring flow rates through a constant-head permeameter in the laboratory.

K_0 was calculated by using Darcy's law:

$$K_0 = \frac{V \Delta Z}{\Delta H}$$

where K_0 = hydraulic conductivity (cm/day); V = rate of flow of water across unit cross-section (cm/day); ΔZ = height interval in the core (cm); ΔH = hydraulic head difference in cm of water.

6.3.3.3. Structural Stability

Structural stability was measured by two methods.

Ratio of Air to Water Permeability (K_a/K_w)

The ratio of air to water permeability was determined as described by Reeve (1953). Aggregates (0·25 to 1·0 mm) prepared from the bulk samples were packed into 5 cm i.d. × 5 cm high cylinders and the flow of air and water measured on the same sample. Intrinsic permeability in each case was calculated using Darcy's law and the viscosity of air and water. The higher the air to water permeability ratio, the less stable the soil.

Modal Pore Size Ratio (r_Q/r_S)

This was determined from the drainage moisture characteristics for quick and slow wetting over the range of 0 to 100 cm suction (Childs, 1940, 1942), where r_Q refers to the pore size after quick wetting, and r_S to the pore size after slow wetting.

6.3.3.4. Exchangeable Cations

Exchangeable cations were determined by extracting soil samples with 1 N NH_4Cl at a pH of 7·0 and analysing the extract using atomic absorption spectroscopy (David, 1960).

6.3.3.5. Bulk Density

Bulk density was measured at 10 cm intervals by taking 360 cm^3 cores from the top soil.

6.3.4. RESULTS

6.3.4.1. Water Entry

The field infiltration, sorptivity and saturated hydraulic conductivity of surface soil are presented in Table 6.11.

Table 6.11. Surface infiltration data for cultivated soil of four soil series on the Central Plain of Thailand

Soil series	Bulk density (g/cm^3)	Initial moisture content by volume	Water entry at 100 min (cm of water)	Sorptivity (cm $min^{-1/2}$)	Hydraulic conductivity (cm/min)
Pb	1·18	0·102	45·0	2·25	0·34
Np	1·27	0·144	8·2	0·68	0·065
Cn	1·26	0·148	5·8	0·53	0·0022
Rb	1·25	0·160	15·6	0·96	0·058

Of the four soils studied, the Pb soil had the highest infiltration capacity. Lateral water movement into the beds was good because sorptivity was high. No waterlogging occurred in the field, which would be expected from the high K_0. On the other hand, the Cn soil had the lowest infiltration capacity. This was a result of low sorptivity and very low K_0 caused by structural instability (Table 6.12). This was aggravated by the presence of a 'hard pan' of extremely low K_0 at 10–20 cm depth which will be described later. The Np soil also had a low infiltration capacity, but was better than the Cn soil because of its higher K_0. The clay Rb soil had a moderate infiltration capacity because of its reasonable sorptivity. The latter reflects the high clay content and better surface structure of this soil. Swelling took place during measurement, making the two-parameter equation described previously inaccurate for this soil.

Profile measurements of K_0 (Table 6.13) showed that for the Pb soil minimum K_0 values occurred at 20–40 cm depth. This corresponded with a maximum bulk

Table 6.12. Saturated hydraulic conductivity K_0 (mm/day) for uncultivated soil of four soil series on the Central Plain of Thailand

Depth (cm)	Soil series Pb	Np	Cn	Rb
Uncultivated soil				
0–10	818	2301	344	44
10–20			1·1	
20–40	32	341	4·7	486
40–60	626	78	2154	558
60–80	2070	10	429	592
80–100	244	6·7	ND	557
Cultivated soil				
0–10	3850	775	42	941

ND, not determined.

density in the profile of 1·65 g/cm³ at 20–30 cm depth (Table 6.13), indicating a definite hard pan layer. For the Cn soil, a minimum K_0 of only 1 mm per day occurred at 10–20 cm depth and K_0 was still low at 20–40 cm depth. This did not correspond with a distinct high bulk density zone in the profile, although there was slightly higher bulk density at 20–30 cm depth. This layer of low K_0 is mechanically strong and a typical hard pan. Such a low value of K_0 would severely limit deep percolation down the profile and is the main cause of waterlogging in the top soil. It is interesting to note that the K_0 below the hard pan is high for this soil.

Table 6.13. Bulk density data (g/cm³) for the four soil series on the Central Plain of Thailand

Depth (cm)	Soil series Pb	Np	Cn	Rb
0–10	1·37	1·42	1·40	1·39
10–20	1·41	1·55	1·45	1·55
20–30	1·65	1·60	1·56	1·61
30–40	1·56	1·62	1·53	1·62
40–50	1·56	1·61	1·56	1·62
50–60	1·60	1·62	1·56	1·66
60–70	1·56	1·65	1·56	1·67
70–80	1·65	1·66	1·56	1·68
80–90	1·63	1·71	1·62	1·68
90–100	1·62	1·72	1·62	1·62

Cultivation Effects

The effect of cultivation on K_0 of the surface soil differed among the soils studied. In the case of Pb and Rb soils, K_0 increased with cultivation and it was noticed that structural breakdown during quick wetting was not severe. The increased porosity resulting from cultivation, as indicated by lower bulk density, was not entirely lost and K_0 was higher as a result. In the case of Np and Cn soils, severe slaking of the actual soil aggregates took place upon soil wetting, destroying the large pores and resulting in low K_0. In the uncultivated condition, the latter two soils possibly had a number of old roots which resulted in an uncultivated K_0 higher than cultivated K_0 (Table 6.5).

Hard Pan Effects

The low flow rate through the hard pan of the Cn soil has been demonstrated in the field by Ponsana and Judd (1969), as shown in Figure 6.2. Tensiometers were installed at 15, 35, and 50 cm depths in a rice paddy that had been flooded for 60

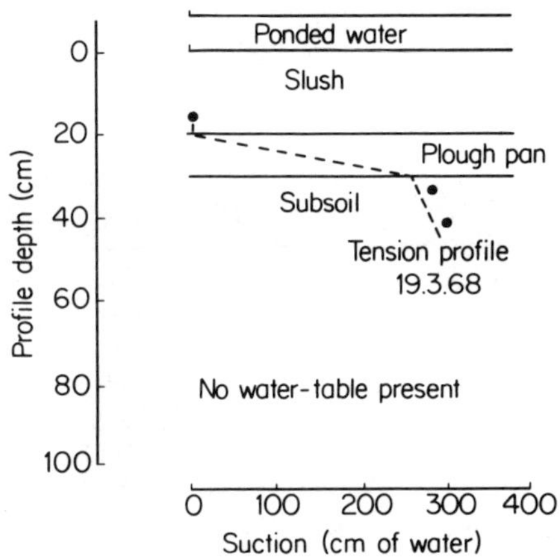

Figure 6.2. Suction profiles in soil growing rice of Cn series

days. Measurements showed that the surface soil was saturated at zero suction but the subsoil beneath the pan had a suction of 300 cm of water. No water-table was detected. The high suction head through the hard pan (30 times gravity) shows that the latter had a very low K_0 and acted as an infiltration throttle.

6.3.4.2. Structural Stability

The structural stability data are presented in Table 6.14. The interpretation of the structural stability data is somewhat difficult because of the different methods used. This emphasizes the problem of subjectivity in measuring structural stability. Taking into account the air to water permeability alone, Reeve (1953) considers that a ratio 100 generally indicates an unstable soil. Using this criterion, the Cn site has the only unstable surface soil. The most stable profile is at the Np site, while the Rb site has increasingly unstable soil below 40 cm. The subsoil at the Pb site below 40 cm depth is extremely unstable, and this is mostly explained by high amounts of exchangeable sodium on the clay fraction (Table 6.15).

Table 6.14. Air to water permeability ratio (K_a/K_w), quick and slow wetting, pore size ratio (r_Q/r_S), for four soil series on the Central Plain of Thailand

Soil depth (cm)	K_a/K_w Soil series Pb	Np	Cn	Rb	r_Q/r_S Soil series Pb	Np	Cn	Rb
0–20	67	30	514	86	0·050	0·112	0·050	0·068
20–40	39	33	342	35	0·057	0·105	0·085	0·092
40–60	28,000	24	317	182	0·102	0·217	0·068	0·137
60–80	123,000	18	409	198	*	0·182	0·060	0·101
80–100	482,000	48	ND	1055	*	0·124	ND	0·107

* Not measurable because of dispersion in the sample.
ND, not determined.

Table 6.15. Exchangeable cations, total cation exchange capacity (CEC), sodium absorption ratio (SAR), exchangeable sodium percentage (ESP) for the four soils

Soil series	Depth (cm)	Exchangeable cations (meq/100 g) Na^+	K^+	Ca^{++}	Mg^{++}	Total CEC	SAR	ESP
	0–20	0·57	1·22	6·83	5·93	38·4	0·23	1·5
	20–40	0·65	0·56	3·13	5·30	28·2	0·32	2·3
Pb	40–60	1·39	0·15	6·88	7·54	51·2	0·51	2·7
	60–80	5·21	0·06	10·00	9·98	32·6	1·65	16·0
	80–100	5·00	0·12	11·00	9·98	30·2	1·54	16·5
	0–20	0·38	0·19	11·25	6·24	45·6	0·13	0·8
	20–40	0·45	0·07	11·25	7·28	44·8	0·15	1·0
Np	40–60	0·20	0·41	14·38	6·66	42·6	0·06	0·5
	60–80	0·26	0·45	12·50	5·93	41·6	0·09	0·6
	80–100	0·52	0·22	11·88	6·97	38·2	0·17	1·4
	0–20	0·33	0·15	11·00	6·24	30·0	0·11	1·1
	20–40	0·75	0·16	9·38	7·70	29·5	0·25	2·5
Cn	40–60	2·89	0·10	6·88	2·78	41·3	1·08	7·0
	60–80	2·76	0·06	8·13	6·66	47·6	1·02	5·8
	0–20	0·24	0·15	20·00	9·57	43·4	0·06	0·6
	20–40	1·37	0·06	12·50	7·90	38·0	0·43	3·6
Rb	40–60	1·55	0·77	12·50	7·59	71·4	0·49	2·2
	60–80	1·37	0·90	11·88	7·59	40·2	0·44	3·4
	80–100	3·33	0·66	12·50	7·90	42·6	1·05	7·8

On the other hand, the quick and slow wetting indicate that all the soils are unstable. The difference between the two sets of data can be explained by the fact that sudden immersion of air-dry aggregates in water during quick wetting is a much more severe treatment than the overnight wetting of the core of aggregates by capillary movement before carrying out the water permeability test. However, the Np soil is the least unstable in this test, which is in agreement with the air to water permeability ratio data..

6.3.5. DISCUSSION

Soil erosion by surface movement is associated with such factors as infiltration capacity and aggregate stability of the surface soil, texture, bulk density, rainfall intensity and duration, and degree and length of the slope. The erosion hazard will be high on soils that have a low structural stability, low infiltration capacity, are subject to rainfall of high intensity and extended duration, and have a steep and long slope.

The infiltration capacity depends in part on the soil type and its physical properties, as shown for soils described above. Sorptivity and hydraulic conductivity measurements can be made to help characterize the water relations of a given soil.

Talsma (1969) found that field soils in Australia had a value of K_0 ranging from about 0·45 to 0·002 cm/min or less for soils with relatively high and low hydraulic conductivity, respectively. Corresponding values of sorptivity range from about 2·3 to 0·5 cm $min^{-1/2}$. Ponding of runoff of water could therefore be expected, depending on the slope and rainfall intensity. The results have shown that increasing the surface roughness (or proportion of large aggregates in the surface soil) by cultivation can also affect the initial infiltration capacity.

With continued heavy rain, any initial increase in infiltration capacity due to cultivation would be expected to decrease as the macropores filled with water, so that in time runoff would again take place. However, the K_0 of the subsoil is sufficiently high (say $K = 0{\cdot}66$ cm/min) if the rain is intermittent and the surface soil has relatively stable aggregates. The infiltration capacity of a soil could be maintained for longer periods before runoff would occur. This applies particularly to soils that are stable under wetting and raindrop action. The present results indicate that if the cultivation of unstable soils, especially if the final proportion of small aggregates (say 25 mm), is high, the infiltration capacity can be decreased because of surface slaking and associated reduction in K_0. Ponsana and Judd (1969) have investigated the fact that in the soil tank model in the laboratory, the rate of water penetration into finely cultivated soil is low. Aggregates of Cn soil were shaped into 75 cm ridges and the wetting front was observed through a glass window in the tank. Infiltration into the ridges was dependent on the height of water maintained in the furrows. Height of water in furrows 3, 6, 9 and 12 cm would wet up the ridges completely in 1670, 1217, 350 and 330 min.

The results have also illustrated the throttle effect on water infiltration down the profile when a zone of low K_0 is present. This can be associated with a naturally occurring subsoil of low permeability or can be created by cultivation, particularly of wet soil as in rice paddy. The puddling of permeable soils within terraces can restrict downward movement of water and increase the ponding effect, thus creating paddy conditions for rice-growing on sloping areas. With such a system of water management, provision would have to be made for the controlled flow of excess water down the slope (i.e. from one terrace to another).

ACKNOWLEDGMENTS

This paper is based on work carried out by the Soil and Water Management Section at the Central Region Agricultural Centre, Chainat, Thailand.

SUMMARY

The hydrologic properties of soils affecting water movement into and down the profile have been measured for four soil series ranging from sandy clay loams to heavy clay in the Central Plain of Thailand. The measurements include field infiltration and sorptivity, vertical saturated hydraulic conductivity, structural stability, bulk density and physico-chemical properties.

REFERENCES

Childs, E. C., 1940, The use of soil moisture characteristics in soil studies, *Soil Sci.*, **50**, 239–252.

Childs, E. C., 1942, Stability of clay soils, *Soil Sci.*, **53**, 79–92.

David, D. J., 1960, The determination of exchangeable sodium, potassium, calcium and magnesium in soils by atomic absorption spectro-photometry, *Analyst*, **85**, 495–503.

Philip, J. R., 1957, The theory of infiltration. 4. Sorptivity and algebraic infiltration equations, *Soil Sci.*, **84**, 257–264.

Ponsana, P., and P. Judd, 1968, Factors affecting water penetration into field ridges, *Annual Report 1968*, Central Region Agricultural Research Centre, Chainat, pp. 9–14.

Ponsana, P., and P. Judd, 1969, Preliminary investigation on water movement using tensiometers, *Annual Report 1969*, Central Region Agricultural Research Centre, Chainat, pp. 14–16.

Reeve, R. C., 1953, A method for determining the stability of soil structure based on air and water permeability measurements, *Soil Sci. Soc. Amer. Proc.*, **17**, 324–329.

Talsma, T., 1969, In site measurement of sorptivity, *Aust. J. Soil Res.*, **7**, 269–276.

Van der Kevie, W., S. Senawongse and P. Pongprot, 1965, Chao Phraya Irrigation Project, Report on the soil survey in the Boromdhart Tract, SSR-39-1965.

PART 7

Farming Systems and Erosion Hazard in Latin America

7.1

Review of Soil Erosion Research in Latin America

R. LAL

There is a growing awareness of the menace of soil erosion amongst various research organizations, agricultural planners and development agencies in Latin America (Conservation Foundation, FAO, 1954a, b, c; Pan American Congress, 1966; Watters, 1971). A survey conducted by the Conservation Foundation, FAO (1954a, b, c) produced an initial map depicting the magnitude of the erosion problem in various countries of Latin America. Any attempt to produce a generalized map can only result in a general guideline because of the diversity of soil and climatic conditions in the region. For instance, rainfall varies from almost nothing in the coastal region of Peru to 5000 mm per year in various regions in the Amazon basin. Because of the rugged terrain and high climatic erosivity, vast areas of highlands in Ecuador, Colombia, and in the Andes have been either destroyed or seriously damaged by erosion (plates 28, 29 and 30). There is little basic research information available on runoff and erosion in Latin America, and hence planning for erosion control is faced with lack of adequate information. A brief review of the available information for some regions is presented below.

7.1.1. CENTRAL AMERICA

7.1.1.1. Mexico

Mexico is one of the countries in the region with a rather serious erosion problem. About 50 per cent of the land of Mexico lies at an elevation above 1000 m, much of it on steep slopes. The Conservation Foundation, FAO (1954a, b, c) reported that hundreds of thousands of acres of land had lost their capacity for production because of serious soil erosion. Since maize is grown over some 50 to 60 per cent of Mexican arable land, there is an urgent need to develop adequate management techniques to control erosion. In some humid forested regions, corn grown with shifting cultivation techniques is giving way to large-scale clearing, causing serious soil erosion (Watters, 1971).

7.1.1.2. El Salvador

The Conservation Foundation, FAO (1954a, b, c) reported that El Salvador, the smallest country of Central America, is faced with an acute problem of land shortage due to population pressure. It is estimated that slight to moderate erosion has taken place on the entire land surface. Intensive cultivation of corn and beans accompanied by deforestation have increased the erosion problem tremendously.

7.1.1.3. Guatemala

The shifting cultivation practised by the native population has given way to commercial plantations, for example coffee, in the highlands, resulting in accelerated erosion (Conservation Foundation, FAO, 1954a, b, c). Steeper slopes have moderate to serious erosion. The lowlands under banana do not suffer from severe erosion hazard. Some earlier work on the soil conditions and erosion hazard under different management techniques has been reported by Cardona and Deger (1934) and Deger (1940).

7.1.1.4. Costa Rica

The overall erosion problem in Costa Rica is not as severe as in Mexico and El Salvador. Ives (1951) reported some basic information on runoff and soil loss from Turrialba. The plots under grass and with a mixed cropping system had less runoff and soil loss than bare plots. Recently Amezquita and Foresythe (1975) have applied the Universal Soil Loss Equation to the soils at Turrialba. The total annual erosivity index R varied considerably from year to year, from the lowest of 70·9 in 1960 to the highest of 318·8 in 1970 and with a mean value of 122 over a period of 15 years. The most erosive months are May to December. The K factor of various soils was reported to range from 0·103 to 0·155.

7.1.1.5. Puerto Rico

The soil erosion in Puerto Rico is primarily confined to the highlands. The total area of this island is 8900 km. Twenty-four per cent of the area has slopes less than 15 per cent, while 22 per cent of the land has slopes greater than 60 per cent. Twenty-five per cent of the land surface is suited only to forest and permanent cover (Hodges and Reichederfer, 1961; Smith and Abruna, 1955). Erosion is therefore a major problem except on flat land.

Bonnet and Lugo-Lopez (1950) reported the relative erosiveness of various soils of Puerto Rico and recommended that steep slopes should be kept under forest or in pastures. These authors determined the infiltration rates of various soils in Puerto Rico. The mean rates varied from 9·5 to 24·1 cm/hr. These high infiltration rates were associated with good aggregation and high contents of free sesquioxides. Smith and Abruna (1955) reported runoff and soil loss from some Puerto Rican red-clay latosols and shallow brown clay immature soils. Soils under natural fallow lost 66 cm of water and 253 tonnes/ha of soil. In comparison, desurfaced fallow plots lost 58·2 cm of water and 303 tonnes/ha of soil over a period of six years. Plots with cultivated crops lost 12·9 cm of water and 36·4 tonnes/ha of soil, and grass fallow lost 25·4 cm of water and 4·4 tonnes/ha of soil. Mulching with cane leaves compared to burning showed few differences in runoff, but a soil loss reduction from 15·2 tonnes/ha to 1·4 tonnes/ha per year. Eroded soil could not be reclaimed even after four years of planting with tropical Kudzu (*Pueraria*). The effectiveness of various cultural practices for runoff and erosion control varied from nil for fallow desurfaced and 15 per cent for fallow normal soil to 98 per cent for sugarcane with complete mulch and for tropical Kudzu.

The erodibility of Puerto Rican soils has been investigated by Lugo-Lopez (1969) and Barnett *et al.* (1970). Lugo-Lopez reported a high correlation (r = 0·79) between the per cent silt and clay aggregated and the erosion ratio. Lugo-Lopez *et al.* (1970) presented data from 750 measurements on field water intake rates for different soils in Puerto Rico. They found a highly significant correlation between the eighth-hour infiltration rate and the preceding one-hour interval rates. Because there was a significant correlation between the third and eighth-hour rates, the former values could be used for predicting infiltration and runoff from a given storm. Barnett *et al.* (1972) reported soil and nutrient loss on Puerto Rican soils using a rain simulator. The final infiltration rates were 3·7, 0·9 and 6·2 cm/hr for Humatas clay, Juneous silty clay, and Pandura sandy loam, respectively. Barnett *et al.* (1970) studied the erosivity (R) index and erodibility (K) factor of various soils and concluded that the Universal Soil Loss Equation cannot be applied in this region, in the form in which it was developed.

7.1.2. SOUTH AMERICA

7.1.2.1. Colombia

Colombia, except for a small area, is a mountainous country with high ranges exceeding 3000 m. The eastern part of the country is a rolling plain with an elevation of less than 300 m. Erosion has been a problem in the areas with heavy cattle populations due to excessive grazing. Erosion is also significant under plantation crops such as cacao, coffee and tobacco. Shifting cultivation practiced on very steep slopes is also an important factor (Plate 31). It is estimated that the total annual soil loss in Colombia is as much as 426 million tonnes, of which clean cultivation is responsible for 80 per cent. It is worth noting that only 2 per cent of the land surface is subjected to cultural practices involving clean cultivation (Conservation Foundation, FAO, 1954a, b, c).

The Association of Agricultural Engineers (1948) estimated that of the 30 million hectares of the Andean region in Colombia, 23 per cent was rocky, 5 per cent was intensely eroded to the point of almost total loss, and 25 per cent was likely to become lost to agriculture within 20 years (by 1970). Twenty and a half per cent of the land surface had already lost a large part of its native productivity, and 24 per cent would have lost it in the near future.

Some good basic research on erosion control was initiated in the coffee plantation region by Suarez De Castro and his colleagues (1949, 1950a, b, 1951a, b, 1952a, b, c, 1955). They reported that losses of soil from uncropped land weeded monthly and from a young coffee plantation were, respectively, 90 and eight times as heavy as those from pastures, and those from a well-shaded old plantation in which plant residue had accumulated were less than from the pasture. Establishment of grass for hay on a contour basis reduced erosion by 98 per cent and runoff by 62 per cent. Suarez De Castro (1952b) reported that average losses of soil in three months were 252·7 and 3·02 tonnes/ha from bare fallow and grassland, respectively. Maize planted immediately afterwards gave a yield 3·74 times as high on the ploughed grassland as on the fallow soil; there was no significant difference in plant composition. In another experiment, Suarez De Castro (1951b) reported that annual loss of runoff and soil, respectively, were 216·4 mm and 11·8 tonnes/ha without contour planting as compared with 87·4 mm and 0·65 tonnes/ha with contour cultivation. In the Chinchina region of Colombia, there was an annual soil loss of 200 tonnes/ha from clean cultivated land of moderate slope. Corresponding losses under a dense crop varied from 1 to 20 tonnes/ha depending on cultural practices and plantation age. Recently, some erosion surveys based on Fournier's climatic coefficient have been initiated by Gomez (1975). Wischmeier's EI_{30} index has also been used by Machado and Machado (1975). There may, however, be problems involved in the applicability of these indices for tropical environments.

7.1.2.2. Ecuador

Various regions in the highlands have suffered from soil erosion, particularly where uncontrolled grazing has been practised on steep slopes. Certain soils have an impervious layer within one metre from the soil surface (Conservation Foundation, FAO, 1954a, b, c). During heavy rains, when the surface soil becomes saturated, landslides are very common. The semi-arid region near Quito is heavily grazed and suffers from serious erosion. The Amazon region, if under forest cover, is protected from erosion. Some practical recommendations for erosion control were outlined by Lizarzaburu (1950). Foglino (1965) reported that large areas of agricultural lands on the Ecuadorian highlands had been destroyed or seriously damaged by erosion. In some cases the remnants of old Inca terraces have caused severe gullying by channelling the water at certain places and releasing large masses of it. The same phenomenon occurs throughout the Andes (Plates 29, 30 and 32a, b, c).

7.1.2.3. Peru

The erosion problem in the high land of Peru, covering 30 per cent of the land surface, is spectacular (Plate 30). Short-term and high-intensity storms are characteristics of the territory. Demographic pressures force people to cultivate hillsides greater than 50 per cent slope (Plate 31). Moderate erosion can be observed throughout the region (Gilardi, 1948). The remnants of terraces under progressive stages of deterioration are still visible (Plate 32). In some places all the surface soil has been washed away, exposing the bedrocks (Plate 33). Estimates of soil erosion in the Andean region have been made by Low (1967) using Fournier's climatic coefficient (Table 7.1). About 50 per cent of the land surface lies in the region of 1000–1500 tonnes/km^2/yr of soil loss. The climatic coefficient of Fournier and Henin (1959), however, underestimates the erosion hazard (see Chapter 2.1). Paulet (1973) has determined the EI_{30} factor for Huancayo and reported it to be low, viz. 41.

Runoff plots have been established in different regions, and useful basic information is being obtained by Felipe-Morales (personal communications).

Table 7.1. Potential erosion in the Andean region of Peru (Low, 1967)

Erosion range (tonne/km^2/yr)	Area (km^2)	Per cent of total area
0–1000	216,040	16·88
1000–1500	670,620	52·38
1500–2000	215,544	16·85
2000–3000	95,400	7·45
3000–4000	34,580	2·70
4000–5000	31,700	2·48
5000–7000	16,120	1·26

The erosion problem in the forest region of Peru becomes serious only when shifting cultivation (Plate 34) is replaced by continuous cultivation, through forest removal. The agricultural expansion of Peru is envisaged in the forest region. It is estimated that 87 per cent of the areas to be cultivated in the future will be in this region. The rainfall in the upper Amazon basin averages more than 3000 mm/an. The effects of different methods of forest clearing on soil productivity have been investigated at Yuramaguas by Seubert (1975). Bulldozer clearing resulted in severe soil compaction, with infiltration rates averaging 10 cm/hr in the burned plots and 0·5 cm/hr in the bulldozed plots.

Some general reports on the magnitude of the erosion problem in Peru are given by Gamarra Dulanto (1946), Gilardi (1948), and Ocampo (1956).

7.1.2.4. Venezuela

There are four major ecological zones in Venezuela: (i) highlands which form the north eastern extension of the Andes, (ii) Maracaibo lowlands, (iii) Orinoco lowlands and (iv) Guiana highlands (Conservation Foundation, FAO, 1954a, b, c).

In the highlands as much as 75 per cent of the cultivable area is found on slopes exceeding 25 per cent. Shifting cultivation is prevalent in this region (Watters, 1971). Most of the cultivated land has been severely or moderately eroded. The erosion problem in the other regions is not serious. Recently Gasperi (1975) has given some quantitative data on the magnitude of the erosion hazard in the Estado Lara region of Venezuela. A further analysis of erosion problems in Venezuela is presented by Pla in Chapter 7.3.

7.1.2.5. Brazil

Rawitscher (1948) reported some data on the water economy of the vegetation of the '*campos cerrados*' in Southern Brazil. The soils are acid red loams and at least 20 m deep. There is no superficial erosion except on overstocked ranges, under bare fallow and coffee plantations, and where uncontrolled burning has been practised.

The areas of the central and western portions of the State of Sao Paulo have a generally undulating to rolling topography and therefore are prone to erosion. In this region basic research on soil erosion and its control has been done by Marques and his colleagues (1961). Marques *et al.* (1961) reported soil losses from annual and perennial crops to vary considerably amongst different soil types (Table 7.2).

Table 7.2. Effect of soil type on soil erosion (tonnes/ha/annum) under different crops (Marques *et al.*, 1961)

Crop	Sandy Bauru	Terra Roxa	Sandy Botu Catu
Annual crops	21·1	9·5	21·5
Perennial crops	1·0	0·6	1·0
Pastures	1·2	2·7	1·2

Pastures and perennial crops had significantly lower soil loss. Bertoni and Pastana (1964) reported data on rainfall, erosion and water losses from four experimental stations in this region. The effects of different tillage and planting methods on runoff and soil loss from the two different soils are shown in Table 7.3. Under similar rainfall, sandy Bauru soil (6 per cent slope, 45 per cent clay) had higher runoff and soil loss than Terra Roxa (10 per cent slope, 15 per cent clay).

Central Brazil consists of soils derived from crystalline rocks, viz. granites, gneiss, and schists (Conservation Foundation, FAO, 1954a, b, c). Some steep slopes predominate in this region. Most of the poorly managed coffee plantations have suffered from degradation due to soil erosion.

Plate 28. Barren hills devoid of vegetation because of severe sheet erosion in the Andes. Runoff from the upper slopes has destroyed old terraces on the lower slopes

Plate 29. Severe gully erosion in the Peruvian highlands

Plate 30. A further example of erosion in the Peruvian highlands

Plate 31. Demographic pressure forces people to practise shifting cultivation on hillsides of greater than 50 per cent slope, usually causing serious erosion

Plate 32. (a) Well-preserved and maintained terraces in the Peruvian highlands. (b) Terraces in the Andes not properly maintained and deteriorating. (c) Remnants of Inca terraces in the Andes

c

Plate 33. Ineffective terraces result in severe erosion and exposure of the bedrock

Plate 34. Small clearings established by shifting cultivation in the Amazon jungle of Peru seldom cause serious erosion if cultivation is not prolonged, and rain forest develops rapidly when the land is abandoned

Plate 35. The tropical rain forest in the Amazon area of Brazil, with clearing made for short-term cultivation

Table 7.3. Erosion losses from two soils in Brazil (Bertoni, 1966a, b)

Treatment		Sandy Bauru		Terra Roxa Mistrada		Corn yield on
Ploughing	Sowing	Soil loss (tonnes/ha/yr)	Runoff (mm)	Soil loss (tonnes/ha/yr)	Runoff (mm)	Terra Roxa (kg/ha)
Down	Down	52·8	106	21·4	64	2579
Contour	Down	42·6	111	14·2	61	2585
Down	Contour	28·2	83	7·0	38	3105
Contour	Contour	19·8	92	4·1	36	3152
Rainfall/yr		1123 mm		1292 mm		

In the past the Amazon region of Brazil has had no erosion problem since it has mostly been under primary forest (Plate 36). However, there are alarming reports of erosion whenever large-scale land development schemes are being introduced. It is hoped that large-scale mechanized clearing and post-clearing cultivation techniques will be undertaken with precautions and with appropriate anti-erosive techniques. Investigations on soil erodibility using a rain simulator have been initiated by EMBRAPA at Parana (Mondardo and Jose, 1975).

REFERENCES

Ahmad, N., and E. Breckner, 1974, Soil erosion on three Tobago soils, *Trop. Agric.* (*Trin*), **51**, 313–324.

Alleyne, E. P. and M. J. Percy, 1966, Runoff and soil loss on two small watersheds in the Northern Range, Trinidad, *Trop. Agric.* (*Trin*), **43**, 323–326.

Amezquita, E., and W. M. Foresythe, 1975, Application de la Ecuacion universal de Perdida de Suelo en Turrialba Costa Rica, *Proceedings Vth Latin Am. Congress of Soil Sci.*, Aug. 1975, Medellin, Colombia.

Associon Den Ingenieros Agronomos, 1948, Agricultural engineers will restore Colombian agriculture, *Agric. Trop. Bogota*, **4**, 8–12.

Auer, V., 1951, Scientific considerations on the conservation of natural resources in Patagonia, *Idia*, **4**, 1–36

Barnett, A. P., J. R. Carreker, F. Abruna, *et al.*, 1972, Soil and nutrient losses in runoff with selected cropping treatments in tropical soils, *Agron. J.*, **64**, 391–395.

Bertoni, J., 1966a, O planto e o preparo do solo em contorno e as perdas por erosao, *I° Congresso Panamericano do Conservacao de Solo*, *Proceedings*, pp, 101–104.

Bertoni, J., 1966b, O planto e o preparo do solo em contorno e o aumentio de producao, *I° Congresso Panamericano do Conservacao do Solo*, *Proceedings*, pp. 393–395.

Bertoni, J., and F. I. Pastana, 1964, Relationship between rainfall and erosion in different soil types of Sao Paulo, *Bragantia*, **23**, 3–11.

Bonnet, J. A., and M. A. Lugo Lopez, 1950, Relative erosiveness of Puerto Rico soil, *J. Agric. Univ. Puerto Rico*, **34**, 294–307.

Bonnet, J. A., and M. A. Lugo-Lopez, 1952, The rate of infiltration of lateritic soils, *J. Agric. Univ. Puerto Rico*, **36**, 161–166.

Cardona, R. O., and Y. E. Deger, 1934, The physical constitution of the soils of Guatemala, *Rev. Agric. Guatemala*, **12**, 1–8.

Ceaar, C. M., and H. Almaida Manfrinato, 1954, The anti-erosive effect of vinesse on the soil, *Rev. Agric. S. Paulo*, **29**, No. 3–4, 75–81.
Conservation Foundation, FAO, 1954a, Soil erosion survey of Latin America, I, *J. Soil Water Conserv.*, **9**, 158–168.
Conservation Foundation, FAO, 1954b, Soil erosion survey of Latin America, III. Agents of soil erosion, *J. Soil Water Conserv.*, **9**, 275–280.
Conservation Foundation, FAO, 1954c, Soil erosion survey at Latin America, II, *J. Soil Water Conserv.*, **9**, 214–219, 223–229, 237.
Deger, E., 1940, General observations on the conservation, fertilization and resources of the soils of Guatemala, *Rev. Agric. Guatemala*, **17**, 305–318.
FAO, 1948, *Soil Conservation. An International Study*, FAO Agric. Studies 4, p. 129.
Foglino, F., 1965, La lotta antierosiva in Ecuador. Introduzione della *Setaria splendida* stapf. per la formazione dei cig looni, *Riv. Agric. Subtrop. Trop.*, **59**, 10–12.
Fournier, F., and S. Henin, 1959, A new proclimatic formula for evaluating the specific degradation of soils, *C. R. Acad. Sci. Paris*, **248**, 1694–1696.
Franco, R. M., 1949, Campaign for soil protection and restoration in coffee growing areas, *Agric. Trop. Bogota*, **5**, No. 5, 39–43.
Gamarra Dulanto, L., 1946, La erosion de los suelos: un problema nacional, *Agronomia*, **47**, 5–12.
Gasperi, R. R., 1975, Evaluacion deil proceso erosivo de las areas planas de la depresion de Quibor, estado Lara, Venezuela, *Vth Latin Am. Congress of Soil Sci.*, Aug. 1975, Medellin, Colombia.
Gilardi, C. A., 1948, La erosion de los suelos por el riego en la costa y Sierra del Peru, CNIEA, Informa 68, La Molina, Lima, Peru.
Gomex, A. A., 1975, Sistema I.U.H. Para determinar el uso y manejo de los suelos de ladera, *Vth Latin Am. Congress of Soil Sci.*, Aug. 1975, Medellin, Colombia.
Grohmann, F., and R. A. Catani, 1949, Impoverishment due to erosion and to cotton cultivation in soil of the Arenito Bauru type, *Bragantia*, **9**, 125–132.
Groenveld, D., 1951, Agricultural possibilities in Peru east of the Andes, *Landbouwk. Tijdschr*, **63**, 813–815.
Hardy, F., 1942, Soil erosion in Trinidad and Tobago, *Trop. Agric.* (*Trin.*), **29**, 29–35.
Hodges, L. H., and F. W. Reichederfer, 1961, Generalised estimates of probable maximum precipitation and rainfall—frequency data for Puerto Rica and Virgin Islands, US Dep. Commerce, Weather Bulletin, Tech. Paper No. 42.
Hoeck, K., 1951, A biological method of measuring the erosive action of wind and water, *Ber. Deut. Bot. Ges.*, **64**, 53–56.
Humbert, R. P., 1951, Erosion as a menace to the sugar industry of Hawaii, *Hawaii Plant. Rec.*, **54**, 55–63
Ives, N. C., 1951, Soil and water runoff studies in a tropical region, *Turrialba*, **1**, 240–244.
Lizarzaburu, B., 1950, How can soil erosion be avoided, *Rev. Divulg. Agric.*, **41**, 55–58.
Low, F. K., 1967, Estimating potential erosion hazard in developing countries, *J. Soil Water Conserv.*, **22** (4), 147–148.
Lozet, J., 1951, Soils of the state of Sao Paulo (Brazil) and their conservation, *Bull. Agric. Congo Belge*, **42**, 422–423.
Lugo-Lopez, M. A., 1969, Prediction of the erosiveness of Puerto Rican soils on a basis of the percentage of silt and clay when aggregated, *J. Agric. Univ. Puerto Rico*, **53**, 187–190.
Lugo-Lopez, M. A., J. Juareg, Jr., and R. Perez-Escoler, 1970, Correlation between the rate of water intake of tropical soils at hourly intervals to the 8th hour, *J. Agric. Univ. Puerto Rico*, **54**, 570–575.
Machado, A. S., and L. F. Machado, 1975, Estudio del indice de erosion fluvial on Colombia, *Vth Latin Am. Congress of Soil Sci.*, Aug. 1975, Medellin, Colombia.
Marques, J., A. Quitilano, J. Bertoni and G. Berreto, 1961, Perdas por erosao no estado de Sao Paulo, *Brangantia*, **120**, 1143–1182.

Molinalli, J. C., 1948, Systematization by terracing of land destined for agriculture, Min. Canad. Agric. Uruguay Dir. Agron. Pub. 933.
Mondardo, A., and M. Jose V, 1975, Encontro nacional sobre pasquisa de erosao com simuladores de chuva, EMBRAPA, Iapar do Parana.
Ocampo, A., 1956, La conservacion del Suelo, *Informaciones y memorias*, **57**, 17–27.
Paulet, M., 1973, Guia para el Planeamiento del uso de les tierras agricoles en les zonas de lluvia del Peru, Universidad Nacional Agraria, La Molina, Dept. de Recrivos de Agra y Suelo, Publicaccion No. 3, 5.
Pedraza, R., 1948, The restoration and conservation of soil fertility, *Agric. Trop. Bogota*, **4**, 49–52.
Pendleton, R. L., 1940a, Soil erosion in the tropics, *J. Forestry*, **38**, 753–762.
Pendleton, R. L., 1940b, Soil erosion as related to land utilization in the humid tropics, *Proc. Sixth Pacific Sci. Congress*, **4**, 905–920.
Pendleton, R. L., 1944, General soil conditions in Central America, *Proc. Soil Sci. Soc. Amer.*, **8**, 403–407.
Prego, A. J., 1951, Soil conservation, *Idia*, **4**, Nos. 42–43, 7–13.
Prieto Bolivar, J., 1951, General view of the problem of erosion in Colombia, *Agric. Trop. Bogota*, **7**, No. 9, 52–54.
Prieto Bolivar, J., 1952, Study of landslide and control, *Bol. Inform.* (*Chinchina*), **3**, Nos. 30, 20, 34.
Ranghel Galindo, A., 1948, Soil conservation in Colombia, *Agric. Trop. Bogota*, **4**, 111–114.
Ranghel Galindo, A., 1952, The erosion problem in Bucaramanga, *Agric. Trop. Bogota*, **8**, 27–29.
Rawitscher, F., 1948, The water economy of the vegetation of the '*Campos Cerrados*' in Southern Brazil, *J. Ecol.*, **36**, 237–268.
Schaufelberger, P., 1952, Practical problems of tropical soil science, *Actatrop*, **9**, 17–51.
Seubert, C. E., 1975, Effect of land clearing methods on crop performance in the Amazon jungle of Peru. M.S. Thesis, N. Carolina State University.
Smith, T. L., 1948a, Land tenure and soil erosion in Colombia, *Proc. Inter.-Amer. Conf. Conserv. Renewable Nat. Resources*, 1948, pp. 155–160.
Smith, T. L., 1948b, Land ownership and erosion in Colombia, *Agric. Trop. Bogota*, **4**, 29–35.
Smith, R. M., and F. Abruna, 1955, Soil and water conservation research in Puerto Rico 1938 to 1947, Puerto Rico Agric. Exp. Sta. Bull. 124.
Suarez, De Castro, F., 1949, Experimental methods in soil and water conservation, *Agric. Trop. Bogota*, **5**, 33–38.
Suarez De Castro, F., 1950a, Relation between rainfall and erosion, *Agric. Trop. Bogota*, **6**, 45–47.
Suarez De Castro, F., 1950b, Investigations on soil conservation, *Agric. Trop. Bogota*, **6**, 51–55.
Suarez De Castro, F., 1951a, Let us practice contour cultivation, *Agric. Trop. Bogota*, **7**, 49–53.
Suarez De Castro, 1951b, Effect of erosion in productivity of soil, *Agric. Trop. Bogota*, **7**, 42–43.
Suarez De Castro., 1952a, Erosion is consuming us, *Agric. Trop. Bogota*, **8**, 27–29.
Suarez De Castro, F., 1952b, Soil loss through erosion, *Agric. Trop. Bogota*, **9**, 35–37.
Suarez De Castro, F., 1952c, Erosion potential of rain inside a coffee plantation and in the open, *Bol. Inform.* (*Chinchina*), **3**, 21–24.
Suarez De Castro, F., 1955, Agricultural land use of Colombia and its influence on soil erosion, *Bol. Inform.* (*Chinchina*), **3**, 24–30.
Tallario, L. A., J. Ipucha Aguerre, C. V. Quevedo *et al.*, 1955, Map of soil erosion in the Pampas region, 1st contribution, North West of the province Eva Peron, *Idia*, No. 88, 21–27.

Watters, R. F., 1971, *Shifting Cultivation in Latin America*, For review paper No. 17, FAO, Rome.

Wood, H. B., 1971, Land use effects of the hydrologic characteristics of some Hawaii soils, *J. Soil Water Conserv.*, **26**, 158–160.

Yamamoto, T., and W. H. Anderson, 1973, Splash erosion related to soil erodibility indexes and other forest soil properties in Hawaii, *Water Res.*, **1973**, 336–345.

Zaffanella, M. J. R., 1951, Relationships between rainfall, organic matter and wind erosion and soil of the Pampas Region, *Idia*, **4**, 37–40.

7.2

Erosion Hazard and Farming Systems in the Caribbean Countries

N. AHMAD

7.2.1. INTRODUCTION

In the West Indies soil erosion by water is the main form of soil loss. The extent to which this takes place depends on climate, vegetation, topography, erodibility of the soils and land-use pattern.

The rainfall of the region is high, ranging from 1270 mm to over 7600 mm per year with intensities ranging from 25 mm–127 mm per hour or more. Rains are frequently accompanied by strong winds. This region is also affected by hurricanes, which generally originate in the Gulf of Mexico. Hurricane storms can cause devastating erosion.

Due to high population density and a relatively long history of settlement, the West Indies is devoid of primary vegetation. As a result of numerous cycles of shifting cultivation and soil deterioration, the secondary growth has degenerated and in many instances consists of slow-growing woods, shrubs, sedges, ferns and coarse grasses.

Topographic effects on soil erosion are as important as the climatic and vegetation effects. Many of the islands are mountainous or hilly, and slopes of 20 to 30 degrees or more are not uncommon.

Many of the soils are highly erodible. In Trinidad, Jamaica and Barbados, agricultural lands of steep slopes and those developed on shales are prone to landslips. Landslips are also common on soils with impervious calcareous clays in the subsurface horizon. Some of the soils developed on shale also form crusts (Ahmad and Roblin, 1971) because of high contents of mica. Under these circumstances erosion is serious even on the relatively flat areas. In some soils of volcanic origin with a cemented impervious ash layer, landslipping is also prevalent. On the island of Tobago, some soils erode at the rate of 1–2 cm of soil per year. It is not surprising, therefore, that in most of the hilly areas of Tobago, the surface layer primarily consists of rock fragments with little or no real soil.

The climate, vegetative cover, topography and soil are all conducive to high soil erosion. Another important factor of soil erosion in the West Indies is the land-use pattern. In much of the Caribbean plantation crops such as sugarcane, cacao, citrus, or coconut or improved pastures are grown on good agricultural land. The peasant farmer cultivates marginal land such as steep hillslopes. Peasant farmers follow slash and burn techniques, and hand-hoe the soil into rows without consideration for the topography. The cultivation cycle varies from one to three years. This type of land use is conducive to erosion. Uncontrolled burning destroys soil organic matter while hoeing damages soil structure and exposes erodible subsoil. The soil is without vegetal cover when the rains begin, causing severe erosion. With a decrease in the length of bush fallow, soil fertility is continuously being depleted by erosion.

In addition to environmental and soil conditions, agricultural practices of the region also lead to severe soil erosion. Frequent landslips, gullying, sheet erosion and a high sediment density in streams and rivers during and after a heavy rainstorm testify to the magnitude of erosion in these islands, as does the shallowness of most of the soils of the islands.

7.2.2. AGRICULTURAL SYSTEMS IN THE CARIBBEAN

Agricultural systems in the Caribbean can be divided into two categories. One is plantation crops on the best lands, the main crops being sugarcane, citrus, cacao, coconut, banana and cultivated pasture. These crops are grown in the flat areas, on deep soils and with a minimum of soil disturbance. Hence soil erosion, though serious, is not an obvious problem. In Trinidad, for instance, the yield of cacao is down to about one-third of what it used to be, and this decline has been partly associated with a gradual loss of topsoil. Erosion has been accelerated by poor stands of cacao caused by *Erythrina* disease. Amongst the plantation crops, erosion losses are more severe in sugarcane. In the West Indies the general practice is to burn the crop prior to harvesting and this leaves the soil exposed for several months before cover is established by the regrowth. During this period, the soil is vulnerable. The same practice of burning the crop prior to harvesting was introduced in Barbados, resulting in an annual loss in yield of about four tonnes of sugarcane per hectare. Three years after, the burning was declared illegal.

Although these crops are grown in large plantations, banana is also grown on smallholdings and occasionally on steep land. If grown with terracing, as is practised in Jamaica, erosion is not as severe as it could be without this anti-erosive technique.

The other system of agriculture is small-scale farming such as gardening and subsistence cropping. In general, this type of farming is practiced as shifting cultivation on sloping land. Some serious soil erosion in the region is associated with it. Brief descriptions of the status of erosion in the various islands associated with or resulting from this type of land use are given below.

7.2.3. STATUS OF SOIL EROSION IN THE ISLANDS

7.2.3.1. Trinidad

A comprehensive investigation of erosion in the island of Trinidad was conducted by Hardy (1942). He reported considerable gullying and sheet erosion in the foothills of the Northern Range, and particularly in the western part with its intensive cultivation. Burning is a common practice for land clearing, landslips are frequent and in some places parent rock is within 15 cm of the soil surface. Although the sandy soils of Las Lomas are highly erodible, erosion is not a severe problem because most of these soils are still under forest. In the Central Range, land creep is a major problem and the southern slopes are steeply scarped and show many landslip scars. Chenery (1952) observed that Brasso clay, the most widespread soil of the Central Range, is severely eroded due to prolonged cultivation up and down the slope. He also reported that the Montserrat clay, an important cocoa soil, is prone to gullying. Both Hardy and Chenery observed severe erosion of the marl and associated red soils of the Naparima District in Southern Trinidad, as indicated by caps of exposed white marl as a common feature of hills of the region. In the Southern Range, with soils derived from sandstone, erosion is observed only if the forest cover has been removed.

Although a systematic soil conservation programme has not yet been organized, subsidies are given by the Government to farmers who undertake conservation measures. Some land in the Northern and Central Ranges has also been bench-terraced.

7.2.3.2. Tobago

Hardy (1942), Brown *et al.* (1965) and Breckner (1971) reported on the status of soil erosion in Tobago. Virtually the entire hilly region of Tobago is susceptible to erosion. These soils are developed from volcanic tuffs, breccias, diorite, and schists. Fortunately, the northern and the wettest end of the island is still under forest or permanent tree crops. Further south, erosion becomes a serious problem. Hardy described heavy erosion in the Castara–Parlatuvier area on the leeward side of the island, where intensive peasant and estate cultivation was being carried out. As a result of shallow surface soil due to severe erosion, vast areas have gone out of cultivation and have reverted to secondary growth.

On the windward side of the island, large patches of volcanic soils are currently being cleared by burning for growing arable crops such as corn, root crops and tomato. There is evidence of sheet erosion, gullying and soil slump throughout this area, even though these soils are comparatively less erodible.

By far the worst eroded area of the whole island is the Mason Hall–Les Coteaux district. Sandy clay loam soils of this area developed on diorite and with irregular and sloping topography are highly erodible. The degradation of these soils started more than one hundred years ago, when they were cleared and used for continuous sugarcane monoculture. With the collapse of the sugarcane industry in Tobago, these lands were managed with the shifting cultivation system. Nowadays, the drier parts of this area resemble desert while the wetter parts are covered only with scrub vegetation. All the same, patches of this land are cleared every year and planted to row crops. Often, in the drier parts, a fire started to clear a small patch of coarse grass for a vegetable garden will spread and the whole hillside will be burned, leaving the soil bare for the torrential rains of the rainy season. Fortunately the parent rock weathers quickly to a friable material which will support perhaps one or two crops every few years. Thus, if the land is not cultivated too often, there will always be some soil left on which crops can be grown, no matter how poor the quality, or how small the return.

Fortunately, erodibility is greatest on these soils which regenerate more quickly than on those developed on volcanic or schist material.

Some cultivators, conscious of soil conservation measures, grow row crops along the contour and observe intercropping of corn (*Zea mays*) and pigeonpeas (*Cajanus cajan*). Some farmers use trash mulch on tobacco planted on steep lands. Burning is used as a clearing technique only for difficult vegetation such as bamboo (*Bambusa* sp.). In spite of all this, horrendous devastation can be seen, particularly in the areas cleared of bamboo by burning.

7.2.3.3. Grenada and Carriacou

According to Vernon *et al.* (1958), Grenada has suffered less from erosion and fertility exhaustion than many other islands regardless of soil, climatic and management problems. One of the reasons is that the island's agriculture is based on two tree crops, cocoa (*Theobroma cacao*) and nutmeg (*Myristica fragrans*). Although neither of these is strictly a soil-conserving crop, erosion is less under tree crops than with annuals. The second reason is that these soils are deep and are formed on basic parent material of high inherent fertility.

In contrast, Carriacou has suffered severe erosion over the entire island, and in many cases there is a total loss of soil down to the bedrock. Skeletal soils in hilly areas developed from ash and agglomerate, soils formed from igneous rocks, and those developed on limestone are all susceptible to erosion and have been severely eroded. This deplorable condition is the result of removing the original vegetation, and of using unsuitable cultivation methods and overgrazing.

Vernon *et al.* make no mention of soil conservation measures or a soil conservation programme being followed in either Grenada or Carriacou.

7.2.3.4. St. Vincent

Soil erosion is a serious problem on St. Vincent (Watson *et al.*, 1958). The three crops which contribute much to the erosion problem are cotton (*Gossypium* sp.), arrowroot (*Maranta arundinacea*) and groundnut (*Arachis hypogaea*). Cultivation of these crops involves soil disturbance and leaving soil bare for a long period. Erosion can, therefore, be seen in all but the flattest areas. Fortunately, the farmers of St. Vincent are erosion-conscious, and the soil conservation measures adopted are of a high standard. Contour farming, strip-cropping, use of grass barriers, drains and silt traps are commonly employed by the cultivators. Unfortunately, fragmentation of the holdings tends to negate the effects of these practices.

7.2.3.5. St. Lucia

Soil erosion is a major problem in St. Lucia (Stark *et al.*, 1966). Some of the eroded lands have been abandoned. Many of the soils of the interior are suceptible to erosion. In areas of allophanic clay soils, slumping and gullying are common. Erosion is still accelerated by clean cultivation on steep lands, which should have been left under their original vegetation. Probably the most severe effects of erosion are apparent on the so-called 'shoal soils' (Vertisols) which contain a layer of indurated material in the substratum. Once the surface soil has been lost it is impossible to grow crops. Regeneration is slow, since these soils occur in the drier parts of the island. Stark *et al.* make no mention of any soil conservation work being carried out in St. Lucia, although some terracing is now being done under the auspices of the Windward Islands Banana Producers' Association.

7.2.3.6. Barbados

Erosion is not a serious problem on the relatively flat coral soils of Barbados. Vernon and Carroll (1967) suggested that erosion control measures may be necessary on the soils of the upland plateau of St. John's Valley. However, erosion is a severe problem in the hilly Scotland District. Cumberbatch (1969) estimated that 70 per cent of the Scotland District is threatened by erosion and that 11 per cent of the soils have reached a severe state of degradation. Landslides and gullying are common throughout this region.

A good deal of work is being done towards reclamation of eroded soils, and an intensive soil conservation programme is being sponsored by the Barbados Ministry of Agriculture. However, land-forming measures adopted for soil conservation have often caused more landslipping and sheet erosion than on the original land. In addition, toxic subsoil is exposed during the reshaping process. Sugarcane has proved to be a valuable crop in stabilizing land both before and after reshaping.

7.2.3.7. Dominica

With the exception of Vertisols and those developed on igneous rocks, the soils of Dominica are more permeable and thus less prone to erosion, in spite of hilly and rugged terrain. Lang (1967) reported that erosion is not serious even on slopes of over 60 degrees under natural vegetation, and over 50 degrees under cultivation. Furthermore, erosion in Dominica is less because most of the land is still under the original forest vegetation. Clearing of this land should be undertaken by the Surinam method (Samson, 1959).

Widespread erosion is reported on the leeward side of the island with soils of vertic characteristics. Poor husbandry has been the main contributive factor to this. Like the soils in St. Lucia, the problem is all the more serious because of the slow regeneration of workable soil material. It appears that there is no organized soil conservation programme in Dominica.

7.2.3.8. Montserrat

Lang (1967) reported that the entire land surface is severely eroded, with no surface soil left intact. Unlike the soils of Dominica, these soils are structurally unstable and highly erodible. Though terracing of some kind is a common feature over much of the island, erosion is still not under control.

7.2.3.9. Antigua and Barbuda

The low rainfall and relatively level topography of most of Antigua and Barbuda have helped to keep erosion in check on these islands (Hill, 1964), though erosion is serious in the hilly areas, particularly where cotton is grown with poor soil management. Monoculture of sugarcane in the past has also added to the loss of much soil in hilly areas. Furthermore, torrential rains often occur after long periods of drought when vegetative cover is sparse and land unprotected. In the drier, northeastern parts of Antigua wind erosion is also a problem.

7.2.3.10. St. Kitts and Nevis

Lang and Carroll (1966) reported that erosion in the mountainous areas of St. Kitts had been checked by reforestation. Accelerated erosion is observed only where forest has been cleared and planted to food crops. This is particularly so for the young soils of the upper glacis. Yield reduction due to erosion is not serious since the subsoil over most of the island is free-draining and contour cultivation is observed. On the island of Nevis, in contrast, much erosion can be seen. In much of the principal agricultural areas, particularly on the smectoid soils, the surface soil has been lost completely. This situation is due primarily to uneconomic farming practices and to the widespread cultivation of cotton.

7.2.3.11. Jamaica

Various workers such as Barker (1963, 1970), Finch (1959, 1961), Morgan and Baker (1963), Price (1959a, b), and Stark (1963, 1964) have indicated that soil erosion is rampant in many parts and on many soils of the island. One of the major factors in this situation is population pressure, which has resulted in an increase in frequency and intensity of land use. The parishes of Westmoreland, Portland, Nannover, Clarendon and St. Elizabeth are especially suffering from extremely accelerated soil loss. The soils developed on limestone, shale and grano-diorite are particularly prone to soil erosion. Soils developed on shale are noted for their shallowness, poor infiltrability and low permeability. These soils are therefore extremely susceptible to soil erosion, and form an important soil group managed by the Christiana and Yallahs Land Authorities.

One of the grave consequences of serious erosion hazard in Jamaica is the disturbance caused in the hydrological balance. Bare hills in the highlands release enormous quantities of water resulting in flash floods in the plains. These flash floods also cause much damage to cultivated lands. On the other hand, water supply becomes inadequate in the dry season because of the reduced water-storage capacities of the catchment.

Morgan and Baker (1963) reported that erosion control was an important preoccupation of the Ministry of Agriculture, and that technical and financial help was being sought from different sources.

7.2.3.12. Haiti

The erosion problem in Haiti is probably the worst of all the islands of the West Indies. The original forest vegetation was removed long ago to make way for shifting cultivation. There have been no erosion control measures. Furthermore, increase in population pressures has resulted in cultivation of marginal lands and decreased the length of fallow period. According to Wood (1963), erosion has destroyed all but a few remnants of such mature soils as might have existed in the mountainous areas. Much of the present land cover is weathered bedrock, and bedrock itself often outcrops on more than 50 per cent of the land surface. In the lower areas, flash floods are common. Much devastation is caused by these, and often infertile material brought down from the eroded mountains is deposited on once productive fields. No measures are being taken in Haiti to remedy this situation and the outlook for the future is not very promising, unless drastic measures are taken to bring the situation under control.

7.2.3.13. The Dominican Republic

Although on the same island as Haiti, the erosion situation in the Dominican Republic is not serious. Natural vegetation cover has been maintained over most of the land surface and the continued existence of large estates prevents extreme fragmentation and consequent degradation of agricultural land. This is not to say that soil erosion is not a problem in the Dominican Republic, nor that the owners of landholdings are conscious of the erosion problem and take measures to counteract it. All the same, an Organization of American States Resources

Survey (Survey of the Natural Resources of the Dominican Republic, 1969) stated that one-half of the land under cultivation in the Dominican Republic was steep, of low productive capacity and high erosion susceptibility. Results of poor land use, besides soil and soil productivity losses, include severe fluctuations in the river discharge resulting in the siltation of major irrigation canals by eroded soil material. Erosion is particularly severe in the Cordillera Central, the Cordillera Occidental, the Cordillera Septentrional and the Cibao Valley. The OAS survey report proposes measures which might be taken to improve the situation and, in fact, in the Dominican Republic, unlike Haiti, there is still some room for hope.

REFERENCES

Ahmad, N., and E. Breckner, 1973, Soil erosion on three Tobago soils, *Trop. Agric.*, (*Trin.*), **51**, 313–324.

Ahmad, N., and J. Roblin, 1971, Crusting of River Soil, Trinidad, and its effect on gaseous diffusion, percolation and seedling emergence, *J. Soil Sci.*, **22**, 23–31.

Barker, G. H., 1963, Soil and Land Use Survey No. 24, Jamaica, Parish of St. Ann, UWI, Trinidad.

Barker, G. H., 1970, Soil and Land Use Survey No. 25, Parish of Trelawny, UWI, Trinidad.

Breckner, E., 1971, Soil erosion in the West Indies and its measurement on three Tobago soils, M.Sc. Thesis, UWI, Trinidad.

Brown, C. B., J. R. F. Hansell, I. D. Hill, J. Stark, and G. W. Smith, 1965, *Land Capability of Trinidad and Tobago No.* 1, Tobago Government Printer, Port of Spain.

Chenery, E. M., 1952, *The Soils of Central Trinidad*, Government Printer, Port of Spain.

Cumberbatch, E. R. St. J., 1966, Soil conservation in the Scotland District, Barbados, *Proc. 1st Pan. Amer. Soil Conser. Cong., Brazil.*

Cumberbatch, E. R. St. J., 1969, Soil properties and the effect of organic mulches on reclamation of Scotland clays, Barbados, *Trop. Agric.*, (*Trin.*), **46**, 17–24.

Finch, T. A., 1959, Soil and Land Use Survey No. 7, Jamaica, Parish of Clarendon, UWI, Trinidad.

Finch, T. A., 1961, Soil and Land Use Survey No. 11, Jamaica, Parish of Portland, UWI, Trinidad.

Hardy, F., 1942, Soil erosion in Trinidad and Tobago, *Trop. Agric.*, (*Trin.*), **29**, 29–35.

Hill, I. D., 1964, Soil and Land Use Survey No. 19, Antigua, UWI, Trinidad.

Lang, D. M., 1967, Soil and Land Use Survey No. 21, Dominica, UWI, Trinidad.

Lang, D. M., and D. M. Carroll, 1966, Soil and Land Use Survey No. 16, St. Kitts and Nevis, UWI, Trinidad.

Morgan, G. W., and R. Baker, 1963, Soil and Land Use Survey No. 23, Jamaica, Parish of St. Thomas, UWI, Trinidad.

Morgan, G. W., 1966, Bench terracing in Jamaica, *Proc. 1st Pan Amer. Soil Conser. Cong., Brazil*, pp. 1–15.

OAS, 1969, *Survey of the Natural Resources of the Dominican Republic*, OAS, Washington, D.C.

Price, R. W., 1959a, Soil and Land Use Survey No. 12, Jamaica, Parish of Hanover, UWI, Trinidad.

Price, R. W., 1959b, Soil and Land Use Surveys No. 8, Jamaica, Parish of Hanover, UWI, Trinidad.

Price, R. W., 1959c, Soil and Land Use Surveys No. 8, Jamaica, Parish of Portland, UWI, Trinidad.

Samson, J. A., 1959, Easier forest clearing, *World Crops*, **11**, 273–275.

Stark, J., 1963, 1964, Soil and Land Use Surveys Nos. 14, 15, and 17, Jamaica, Parishes of St. Elizabeth, Westmoreland and Manchester, respectively, UWI, Trinidad.

Stark, J., P. Lajoie, and A. J. Green, 1966, Soil and Land Use Survey No. 20, St. Lucia, UWI, Trinidad.

Vernon, K. C., H. Payne, and J. Spector, 1958, Soil and Land Use Survey No. 9, Grenada and Carriacou, UWI, Trinidad.

Vernon, K. C., and D. M. Carroll, 1967, Soil and Land Use Survey No. 18, Barbados, UWI, Trinidad.

Watson, J. P., J. Spector, and T. A. Jones, 1958, Soil and Land Use Survey No. 3, St. Vincent, UWI, Trinidad.

Wood, H. A., 1963, *Northern Haiti: Land, Land Use and Settlement*, Univ. of Toronto Press, Toronto.

7.3

Soil Erosion in the Humid Tropics of Latin America with Particular Reference to Venezuela

I. PLA

7.3.1. INTRODUCTION

As in any other part of the world, the problems of soil conservation in the humid tropics of Latin America are mainly determined by the soil physical properties, the topography of the area, the rainfall characteristics, the type and kind of agriculture and the technological level reached by that agriculture. It is the combination of these factors that has to be considered to foresee the kind and degree of the problems, as well as select the most appropriate solutions for them. To forget some of those factors may lead to solutions that are not effective or that cannot be applied under practical conditions. The relative importance of each will vary according to the agricultural areas under consideration. If we attempt to analyse how these factors vary in the humid tropical areas of Latin America, it rapidly becomes apparent that it is very difficult to describe general solutions for them. We shall not consider here the problems of soil conservation and management in the highlands of the Andes, not because they are not important, but because they do not belong to the humid tropics, even if they are located in tropical latitudes. It was, however, in these areas that the first problems of soil erosion were detected and management practices to prevent them were applied, even before the European colonization of Latin America. Nowadays the main interest in soil conservation is still concentrated on those areas. In the lowlands of the humid tropics of Latin America, the extensive agricultural development is relatively recent, and other problems of soil conservation besides erosion, like soil compaction and physical degradation in general, have to be considered in relation to management practices. This short report will be mostly dedicated to the consideration of some of the problems of soil conservation prevailing in the main agricultural areas of Venezuela, which are probably similar to those of other tropical countries in Latin America. These areas are generally at 100–400 m above sea-level, with annual precipitation of 900–1600 mm. More humid areas in the south of the country are not seen at the present time as suitable for intensive agricultural development. The examples presented in the different figures are for soils of the Western Plains of Venezuela, the most important agricultural area of the country.

7.3.2. CLIMATIC CONDITIONS

The climate, as in most of the sub-humid tropical areas of the world, is characterized by a high uniform temperature during the whole year, high radiation, and a very definite seasonal distribution of rainfall, with one or two rainy periods followed by a completely dry season of 4–7 months, with strong variability from one year to another. During the wet season, most of the rainfall is concentrated in a few days, and even in a few hours, with frequent dry spells. The rainfall intensity, especially at the initial and final periods of the rainy season, reaches very high values, for variable periods of time. Potential evapotranspiration rates are high, even during the rainy season, because of little cloud cover in the dry periods between rainfall (see Figures 7.1 and 7.2).

7.3.3. SOIL PROPERTIES

Most of the soils used for intensive agriculture in the area belong to the orders Entisols, Inceptisols and Alfisols, with some Ultisols and Vertisols in the marginal higher and lower parts. The Entisols and Inceptisols are more

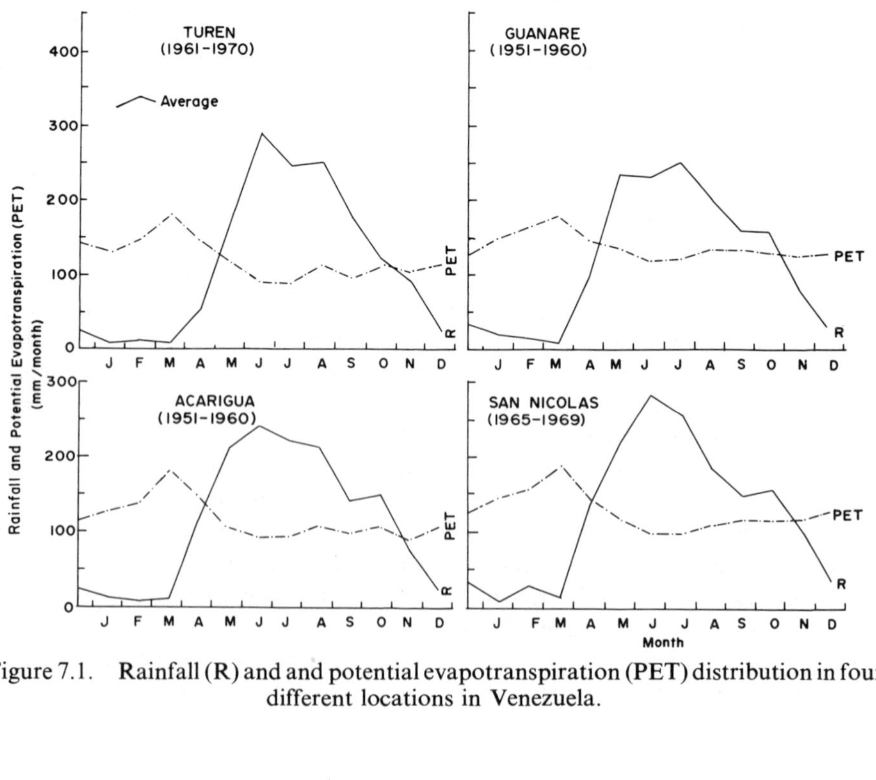

Figure 7.1. Rainfall (R) and and potential evapotranspiration (PET) distribution in four different locations in Venezuela.

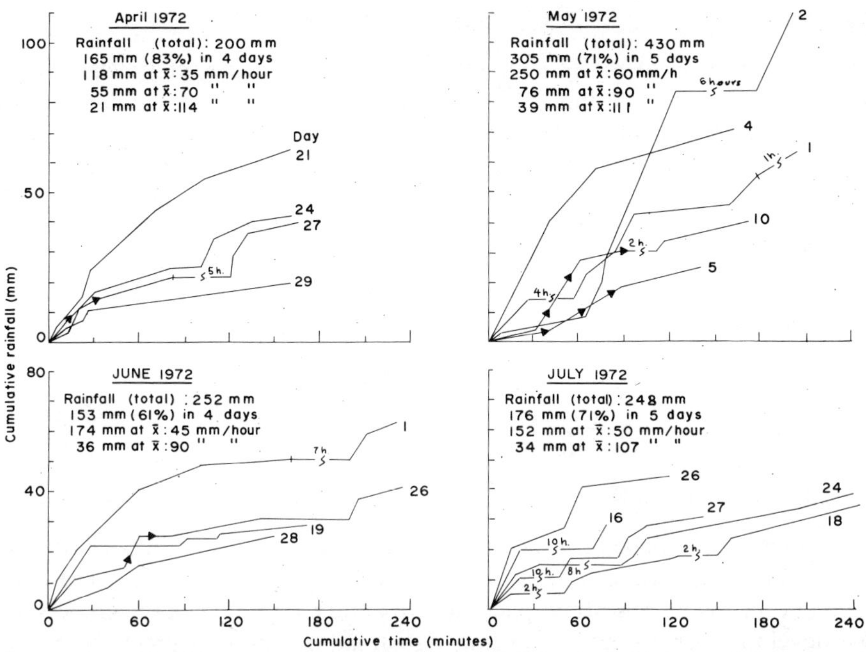

Figure 7.2. Distribution and intensities of rainfall in the initial months of the rainy season (location: Acarigua)

commonly found in the flatter and relatively lower positions in the landscape, with Alfisols dominating in the undulating and rolling surfaces occupying higher positions. There is a high variability in texture of the surface soil, and in internal and external drainage conditions. A common property is the very low stability of the soil aggregates when they are submitted to direct raindrop impact. That property is a consequence of the low cohesivity among soil particles mainly composed of silt and fine sand with variable amounts of kaolinitic clay (1:1 type) and aluminous chlorites (2:1 type with reduced expandibility because of Al hydrous oxide interstratifications). The rapid decrease in organic matter content, after the soils are put under intensive cultivation, also contributes to the low stability of the aggregates. Another consequence of the particle size distribution and low cohesion commonly found in these soils is the tendency to form compact layers easily.

7.3.4. TOPOGRAPHY

Generally, the soils developed in alluvial deposits in the higher parts of the Eastern and Western Plains of Venezuela, and in several valleys and local basins elsewhere in the country, are preferred for intensive agricultural development. The topography is generally flat or with gentle slopes. Extensive agriculture, except for some slope-protecting crops like shaded coffee, is not commonly developed in the lands with high slopes in the humid tropical areas of Venezuela.

7.3.5. AGRICULTURE

With the exception of many scattered areas around the country, where some mixed cropping and shifting cultivation is practised, most of the agricultural production is based on a continuous rotation with seasonal crops like corn, sorghum, beans, sesame, peanuts, cotton and vegetables, or a continuous cultivation of annual or permanent crops like sugarcane, bananas, plantains, cassava, cocoa, coffee and pastures, with some irrigation in the dry season (Figure 7.1). The use of heavy machinery for agricultural operations has become a common practice in most of the areas, not only in large-scale enterprises but also in the small family plots of new colonizations where heavy equipment is used in a cooperative way.

7.3.6. PROBLEMS OF SOIL CONSERVATION AND MANAGEMENT

The main problems of soil conservation and management are associated with the agricultural production of seasonal crops. With some annual crops like sugarcane, problems of soil compaction and salinity, especially of the saline-sodic type, develop in some areas. According to their nature, the main problems may be classified as: (a) compaction, (b) crusting, (c) erosion.

Such problems are not independent, and are often present at the same time, or one of them may lead to the others. After a period of years, complete degradation of the surface soil may occur.

Compaction is a common problem in most of the areas with seasonal or annual crops. The sharp decrease in the interaggregate pores generally leads to aeration deficits, mechanical impedance of root growth, decrease in water penetration and storage, and increase of runoff in sloping lands. Sometimes the problem affects the whole surface soil, but in other cases it is more manifest in a plough layer developed under intensive mechanized agriculture. With almost no cracking, because of the reduced expansibility of the clay fraction, the effects of compaction are much more severe. When the problem of compaction is already present, sometimes even under natural conditions, the beneficial effects of tillage operations are only temporary, especially if the soil surface remains uncovered under intense rainfall.

Among the main reasons for the compaction problem we may mention:

(1) Mechanical composition of the surface soil, with high proportion and continuity in the size fractions of silt (mainly micaceous) and fine sand. The low cohesion among these particles, and their shape, favours their orientation for close packing.

(2) Mechanical effects of raindrop impact, flattening and compacting the loose surface layer of soil which develops with tillage operations.

(3) Use of heavy machinery, often under inappropriate moisture conditions, for tillage practices. As the degradation of the soil physical conditions progresses,

more and more tillage operations are required to keep the surface soil loose, which further aggravates the problem.

(4) Intensive use of the soil for seasonal crops with no or very reduced surface covering during critical periods.

Crusting in many soils used for intensive agriculture in Venezuela is very important, because of the marked direct effects on seedling emergence and water penetration, and also because of the long-term effects on the general physical degradation of the whole surface soil. The problem is generally milder in the Alfisols and Ultisols occupying relatively higher positions in the landscape, and in the sandy Entisols and Inceptisols of the Eastern Plains.

Crust formation is generally associated with:

(1) The low water stability of the agrregates, and the high tendency of the silt and fine sand fractions to be oriented in the process of sedimentation following suspension in rainwater.

(2) High intensities and kinetic energy of the rainfall.

(3) Crop rotations which keep the soil surface uncovered at the beginning and at the end of the dry season.

The main direct consequences of crusting are the sealing of the soil surface, with sharp decreases in water penetration and storage (Figure 7.3), in critical periods of the crop before and after emergence. In flat lands the excess of water

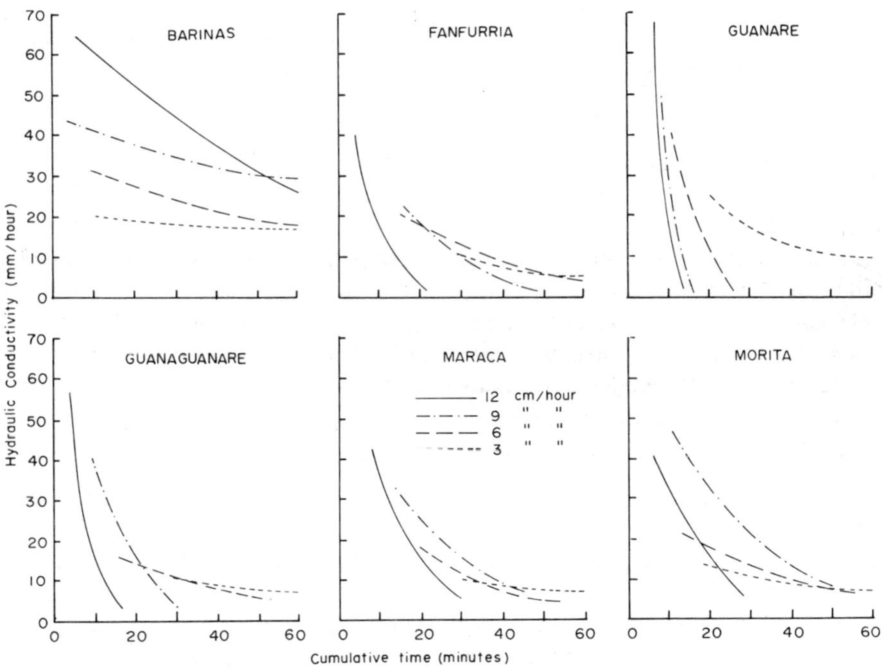

Figure 7.3. Hydraulic conductivity changes with time of surface soil under different rainfall intensities

remains on the sealed surface and creates anaerobic conditions for periods of time that may be critical for the crop. If that problem is prevented by natural or artificial surface drainage, the reduced amount of water stored in the soil may not be sufficient to provide the crop requirements during the very common dry spells at the beginning of the rainy season. In naturally sloping lands, or in lands where the relief has been modified to improve surface drainage, the increase in runoff may lead to high losses of soil by sheet erosion.

In soils with higher clay contents, the resistance of the crust after it dries out reaches values sometimes higher than 4 kg/cm^2, which in many cases prevents the emergence of seedlings.

A general consequence of these problems is the nearly complete loss of the crop, and a requirement for repeated tillage, which has to be done under inappropriate soil moisture conditions. Other consequences are retardation of the crop, affecting the normal rotation with the following crop, and increase in costs of soil management because of the repeated tillage operations.

Erosion is becoming more important as the intense agricultural practices with seasonal crops are extended to the better-drained soils, mainly Alfisols, with undulating surfaces. It is also increasing in the flat lowlands, where the modification of the micro-relief to solve the problems of surface drainage has enhanced the transporation of the easily detached soil material by the runoff water. In both cases, the intensification of agricultural practices and continuous rotation, with clean cultivation and almost no incorporation of crop residues, has also led to an intensification of the erosion processes.

Among the main factors leading to soil erosion in the agricultural lowlands of Venezuela, we may mention:

(1) High detachability of the soil particles, derived from the previously cited soil physical characteristics.

(2) High erosivity of the rainfall, associated with the high intensities and distribution.

(3) Low infiltration rates particularly associated with crust formation and sealing of the soil surface.

(4) Exposure of uncovered soil to raindrop impact derived from the commonly used rotations of seasonal crops and clean cultivation practices.

(5) Extension of intensive agriculture to lands with undulating or sloping surface, and modifications of the microrelief in flat lowlands to improve surface drainage.

The problems of sheet erosion in the humid lowlands of Venezuela have not received the same attention as the more evident problems of gully erosion in the highlands and slopes of the Andes region. There are several reasons for this. One is that the losses of soil by progressive sheet erosion are not so evident as gully erosion developed in sloping lands. Another reason is that most attention has been given to solving problems of surface drainage, considered to be the main limiting factor for agricultural production in these areas. Finally, the use of better-drained, but also less fertile, soils for intensive agriculture has only been extensively developed in the last decade, when the use of larger doses of fertilizers has become a more common practice.

7.3.7. MANAGEMENT PRACTICES

Amelioration of soil compaction and sheet erosion has been tried in isolated cases by means of minimum tillage. Compaction limits crop production, and this has not been very successful. Problems of crusting, compaction and soil erosion are avoided in many cases by the planting of permanent crops or pastures in the most difficult soils. The use of mulches has not been considered a possible practice for large-scale agricultural production of seasonal crops. Leaving crop residues on the soil surface is being considered, but is not extensively used. Irrigation to provide supplementary water in order to avoid the complete dependence of the crops on the very irregular rainfall distribution has been going on for the last 20 years, but the use of irrigation as a tool for changing to crop rotations more convenient for soil conservation has been very limited. Some studies have been done in the last few years to characterize the erosivity of the rainfall and to determine the susceptibility of different agricultural soils to compaction, crusting, and erosion. Studies on the use of soil conditioners, mainly bitumen emulsions, for preventing crust formation have been initiated, with very promising results.

REFERENCES

Dumith, D., 1969, Algunas consideraciones sobre los suelos perados en Venezuela, *Fitotecnia Latinoamericana*, **6**, 235–248.

Homes, P., 1968, Estudio agrologico detallads sector Bocono-Maspavro, Edo Barinas, Guanare, Division de Edafologia (MOP).

Officina Edafologia de Occidente, 1968, Estudio agrologico detallads sistema de riego rio Guanare, Guanare, Division de Edafologia (MOP).

Pla, I., and G. Campero, 1971, Algunas propriedades destructurals de suelos de los llanos altos occidentales de Venezuela y su relacion con ciertas caracteristicas, *Agron. Trop.*, **21**, 433–447.

Pla, I., 1975, Effects of bitumen emulsion and polyacrylamide on some physical properties of Venezuelan soils, '*Soil Conditioners*', Soil Sci. Soc. Amer, Special Publication No. 4, 35–47.

Pla, I., G. Campero and R. Useche, 1974, Physical degradation of agricultural soils in the Western Plains of Venezuela, *Trans 10th Int. Congress of Soil Sci., Moscow*, **1**, 231–240.

Zink, A., and P. Stagno, 1966, Estudio edafologvo de la zona Sto. Domingo-Pauey, Edo Barinas, Guanare, Division de Edafologva (MOP).

PART 8

Towards Erosion Control in the Humid Tropics

8.1

Soil Erosion in the Humid Tropics: The Need for Action and the Need for Research

D. J. GREENLAND and R. LAL

8.1.1. INTRODUCTION

From the previous chapters of this book two facts become abundantly clear: (i) in the humid tropics soil erosion is a serious problem, certain to become a greater threat in the immediate future as the need to use soils for intensified food production expands, (ii) our present knowledge of the many factors determining the amount of erosion occurring in these areas on different soil types and under different soil management methods is at present inadequate to design secure, economic and readily applicable control systems.

While there is a need for further research, there is a need for immediate action, or we may discover the best methods for erosion control after the removal of topsoil and the creation of gullies has produced an infertile desert from the potentially productive lands of the tropical forests.

There are several basic principles which can be used to guide immediate action in developing the humid tropics for increased agricultural production. The most important of these is to maintain a vegetative or mulch cover over the soil at all times to protect it from the erosive force of the rain. This is a universal rule, though it may not be of the same critical importance in the comparatively limited areas of deep fertile soils found on basic rocks in areas of relatively level topography, or on some of the highly acid soils of very stable structure. For most of the humid tropics it is of critical importance. If a cover is maintained, soil loss can be negligible; if it is not, soil losses can be catastrophic. The most satisfactory way of maintaining a cover over the soil is to restrict its use to forestry, or to perennial tree crops or pastures. Provided the establishment of a perennial crop is done with sufficient care, no serious ersosion should ensue.

Localized population pressures make the production of food crops in many parts of the humid tropics an economic necessity. There are also areas, as in Central America, where social constraints make it essential for sections of the community to derive their subsistence by producing food crops in areas of high erosion hazard. In these areas it is essential that governments should be made aware of the seriousness of the problem, and that responsible soil scientists should bring the problem to their attention with adequate conviction and force of argument, so that the governments not only pass the necessary legislation to control agricultural activities in these areas, but develop the will to see that the necessary measures are enforced. Although immediate social disruption can arise through erosion control measures, the long-term benefits far outweigh immediate difficulties.

An even greater danger arises in many of the less populated parts of the humid tropics, when well-intentioned 'development schemes' are prepared to introduce 'modern' agriculture into the forest. In many parts of the humid tropics large machinery has been used to scrape away the forest and litter, and with it the few inches of fertile topsoil, crop production schemes have been introduced without adequate erosion control measures, and poor and declining production has resulted—blamed usually on poor management or poor labour quality or any factor other than inadequate attention to the need to conserve the soil and soil fertility.

In such instances the first need is for sufficient efforts to be made to ensure that those responsible for such schemes are aware of the attendant dangers of soil erosion. A catalogue of the unsuccessful attempts to introduce large-scale mechanized arable agriculture to the humid tropics would be daunting to those involved. Governments and commercial concerns are naturally not anxious for their failures to be publicized. The United Nations Environment Programme and Food and Agriculture Organisation are jointly planning a world survey of soil degradation (FAO, 1975). This is undoubtedly a step in the right direction, and will give some needed publicity to the soil erosion problem and help to mark those areas where the problem is currently most serious. It should also provide factual information on the extent of the problem, so that economic assessment of its importance can be put in quantitative terms. This is undoubtedly desirable, but must not be thought of as a substitute for actions to prevent the erosion occurring now. It is already well established that soil erosion is a major problem of agricultural development of the humid tropics.

Certainly every agricultural development scheme involving large-scale clearing operations should be assessed in terms of erosion risk before it is undertaken, and the necessary control methods should be included as part of the essential costs of any development scheme. These have to be based on existing knowledge, and so are liable to be less than perfect. At present, adherence to agronomic rather than engineering control methods provides much the safer course, since it is in relation to engineering design criteria that good data on control methods are most lacking (Plate 36).

8.1.2. RESEARCH NEEDS

To design appropriate soil management systems for the humid tropics, and appropriate methods for erosion control, it is essential to have a clear understanding of the causes of erosion, and quantitative information on the factors involved. The so-called Universal Soil Loss Equation of Wischmeier and Smith (1960) certainly provides the framework for a quantitative analysis of the problem. As pointed out in earlier chapters it is not truly universal, and the factors need to be carefully defined and determined for its use in the humid tropics.

8.1.3. EROSIVITY

The erosivity of tropical rainstorms is not conveniently defined by the 'EI_{30}' index. Other more appropriate and simpler indices are under study, as described in Part 3. It is already apparent that rainfall in the lowland humid tropics is more erosive even than that of areas such as the southeastern and midwest United States, where the seriousness of the soil erosion problem has long been recognized.

An immediate research need for the humid tropics is to obtain much more data on climatic erosivity, so that appropriate iso-erodent maps can be prepared to guide agricultural practice throughout the area. The present scattered records are far from sufficient to produce useful maps. An increase in erosivity with increasing total rainfall, and with decreasing elevation, seems to occur, but more data are required to establish this, and the quantitative magnitude of the erosivity. Again, there is need for critical evaluation of much new information to ensure that the most appropriate erosivity index is selected to guide erosion control practices.

8.1.4. ERODIBILITY

It is not easy to define erodibility as a completely intrinsic soil property. The two factors 'detachability' and 'transportability' are essential components, but both can vary in time for any one soil type, as cultivation and cropping alter the surface structure and organic matter content of the soil. With these the porosity due to biological activity, and the cohesiveness associated with particle disposition, also change. The rate of rainfall acceptance depends not only on the force and rate at which it arrives at the soil surface, but also on the ability of the soil to accept it, and allow transport within the profile. The transport may be vertical or lateral depending on the conductivity in each direction. Frequently the conductivities differ markedly between horizons. Antecedent moisture contents of profiles will also determine rainfall acceptance, and therefore likelihood of incomplete rainfall acceptance.

The problems associated with definition of erodibility, the factors determining structural characteristics, and methods for field determination of erodibility values have all been discussed earlier. The most satisfactory data are obtained from standard runoff plots where erosion is measured on a specific soil type over several years. However, the number of soils where such plots can be sited, and the data collected with necessary attention to detail over several years, are very limited.

The necessary information for a wider range of soils must be collected using mobile rainfall simulators which can be taken to different sites and erodibility determined under different conditions. Although a start has been made in this direction (Part 3) very many more data remain to be collected. Such information can provide relative erodibility values. It is, however, essential that sufficient runoff plots are established to enable these data to be used in the context of the soil loss equation.

Information on erosivity and erodibility is of course necessary to guide studies of structural properties of soils. The importance of interparticle forces in determining structural stability can only be assessed if the magnitude of the stress to which the aggregates are likely to be subjected is known. Much more information is required to understand how intrinsic soil properties determine both detachability of soil particles under raindrop impact as well as the size of the particles, usually microaggregates of silt size, which are subject to transportation in runoff water. At present, the relationship between soil type and erodibility is very poorly defined. The classification of tropical soils is still imperfect. Although the American system (USDA, 1975) and FAO legend to the Soil Map of the World (FAO, 1970) both represent important advances, neither can claim to include erodibility as an important criterion. Similarly, the lack of well-defined erodibility values for the major soils of the humid tropics presents difficulties in developing valid land-capability classifications. A combination of experimental studies with the information to be compiled by the UNEP/FAO survey of soil degradation should enable a much surer basis to be developed for classifying soils in terms of their liability to erosion, and hence in placing soils in appropriate land-capability classes.

8.1.4.1. Land-form Factors

The effect of length of slope, slope form and angle of slope on soil erosion have been little studied in the humid tropics. Again, use of rainfall simulators should produce much valuable information, essential to the development of properly constructed engineering designs for runoff control. The work discussed in earlier chapters and recent publications (Lal, 1976) shows that much has still to be learnt of the differences which occur between results obtained from temperate areas and those found in the humid tropics.

8.1.5. MANAGEMENT

Throughout this book the theme has been pursued of how to develop more intensive food crop production systems for the humid tropics without leading to increased erosion. The fact that even with traditional erosion control measures,

Plate 36. Well constructed erosion control systems based on inadequate design criteria frequently fail and lead to serious damage. Lack of adequate design criteria is a major deficiency in the tropics. An example of failure from Kongwa, Tanzania. (Photo courtesy Prof. N. Hudson and Aerofilms Ltd.)

[*Facing page* 264

such as contour ploughing and water control bunds, exposure of the soil to the extreme erosivity of the climate is likely to cause accelerated erosion, has been emphasized many times. Methods of cultivation which maintain a vegetative cover over the soil have been shown to be the most successful methods of minimizing damage. This can mean use of perennial crops, mixed and relay cropping to keep some plants in the ground all the year, strip cultivation techniques with only the seed line cultivated, or with wider cultivated strips alternated with grassed strips, or, perhaps potentially the most useful, zero tillage methods using herbicides for weed control and residues of crops and dead weeds to maintain a permanent mulch.

There will also be areas where properly designed engineering methods of erosion control are appropriate, economic and successful. Research on the necessary design criteria is essential. Equally, many studies are needed of the various agronomic production systems which are compatible with increased production and erosion control.

REFERENCES

FAO, 1970, Key to the soil units for the Soil Map of the World, FAO, Rome.

FAO, 1975, A world assessment of soil degradation—an international program of soil conservation, Report of an expert consultation on soil degradation, Rome, 10–14 June, 1974.

Lal, R., 1976, Soil erosion on Alfisols in Western Nigeria. I & III, *Geoderma*, **16**, 363–376 and 389–402.

USDA, 1975, *Soil Taxonomy*, USDA, Washington.

Wischmeier, W. H., and D. D. Smith, 1960, A universal soil-loss estimating equation to guide conservation farm planning, *7th Intern. Cong. Soil Sci.*, **1**, 418–425.

Appendix

Recommendations of the Conference
Common Units and Conversion Tables

RECOMMENDATIONS OF THE CONFERENCE ON SOIL CONSERVATION AND MANAGEMENT
June 30–July 4, 1975 (IITA, Ibadan)

In view of the extreme erosivity of the climate of the humid tropics, production of food crops in these areas is an operation which can lead to loss of topsoil and further erosion problems, decreasing the ability of the land to produce food crops. Nevertheless, the increasing populations of these areas demand that food production is increased, and in some areas very substantially. A coordinated research effort is therefore recommended, to establish systems of increased food crop production not attended by increased erosion hazard. Therefore:

(1) Zero and minimum tillage techniques, with mulches, should be examined in a wide range of environments, with herbicides and appropriate ULV sprays as a possible accompanying technique, as a basis for the development of more productive farming systems.

(2) Careful studies of soil erosion and other factors should be undertaken when any intensification schemes for agricultural production are undertaken.

(3) As a base guideline for the development and application of improved farming practices, it is necessary to develop erodibility and erosivity indices for the major climatic and soil regions of the area. To facilitate the acquisition of this information, and to coordinate the preparation of regional maps, an international network is proposed. It is recommended that funds be sought to enable the regional coordinators of this network to meet at appropriate intervals, and the report to a Second Workshop on Soil Erosion and Farming Systems in the Humid Tropics, to be convened at the International Institute of Tropical Agriculture, Ibadan, in or near July 1977.

(4) Studies of soil erosion should be integrated within studies of catchment hydrology, and the effects of changes in soil management on the hydrology of those catchments. Further catchment studies in different areas of the humid tropics are required. These should include as an integral part water budget determinations.

(5) It is understood that a programme of assessment of soil degradation on a worldwide basis will be undertaken by FAO, WMO, UNEP, UNESCO and UNDP in collaboration with other bodies. Coordination of this work with that recommended here is necessary and suitable collaborative arrangements should be made.

(6) It is recommended to all governments in the humid tropics that appropriate attention be given to the serious problems posed by the increased

hazard of soil erosion associated with more intensive agricultural practices. The importance of adopting farming systems which are appropriate to the soil type and land form and so less likely to cause erosion is emphasized. Research on this problem should be intensified.

(7) It is recommended that training courses be organized for scientific and extension workers in the areas of modern aspects of soil management and erosion control practices. These training courses could be organized at IITA in collaboration with the national training schools being proposed in various countries in the region.

(8) It is recommended that climatic data relevant to soil erosion problems be collected by national meteorological services and made available to interested workers.

(9) It is recommended that the preparation and distribution of a regular newsletter be undertaken by the International Institute of Tropical Agriculture. The newsletter should contain information on progress in soil erosion research, and on meetings, courses and projects concerned with soil erosion.

Terms of Reference for an International Network

(a) To collate existing information from research stations concerning:

(i) daily records of rainfall amount;
(ii) daily records from recording raingauges;
(iii) wind velocity records;
(iv) erodibility.

These records should be used to compute existing indices, for example EI_{30}, $KE > 1$ and AI, from which iso-erodent maps may be prepared.

(b) To recommend experiments to be initiated on the measurement of the erosivity of rainstorms in selected environments and at different altitudes. Measurements of rainfall characteristics such as drop size distribution and kinetic energy should be encouraged at selected stations, and of the wind factor in relation to erosivity.

(c) To encourage measurements of erosivity at a large number of stations, using Ellison-type splash cups with standard sand.

(d) To establish which soil parameters are most closely related to the erodibility of soils of these areas, to encourage the appropriate research needed to determine this, and to develop suitable recommendations regarding land use and land capability maps.

(e) To recommend for which soils it is most important to collect further erodibility data, and to make recommendations to see that the appropriate experiments are conducted.

(f) To recommend such further measures as may be needed to ensure that the effects of topographical factors on soil erosion are appropriately evaluated.

(g) To develop, along with the R (erosivity) and K (erodibility) factors, a first approximation of the C (crop management) and P (erosion control practice) factors for an erosion equation. This will include approximation of the effects on soil erosion of various methods of clearing, cultivation, weeding, fallow, relaying of crops (including sole cropping), harvesting, and the effectiveness of various mechanical practices. After this first approximation is attempted, the working party should make recommendations for collecting data needed for refining the *C* and *P* factors.

COMMON UNITS AND CONVERSION TABLES

Length

1 inch = 25·4 mm
1 foot = 0·3048 m
1 yard = 0·9144 m
1 chain = 20·12 m
1 mile = 1·609 km

Volume

1 ft^3 = 0·02832 m^3
1 yd^3 = 0·7645 m^3
1 gal (UK) = 4·546 litres
= 0·004546 m^3
1 gal (US) = 3·785 litres
(= 0·003785 m^3
1 ft^3 = 6·23 gal (UK)
= 7·48 gal (US)
= 28·32 litres
1 $mile^3$ = 4·166 km^3
1 acre-ft = 271,400 gal (UK)
= 1613 yd^3
= 1233·5 m^3
1 million gal (UK) = 3·7 acre-ft
= 4564 m^3

Area

1 in^2 = 645 mm^2
1 ft^2 = 0·0929 m^2
1 yd^2 = 0·8361 m^2
1 acre = 4047 m^2
= 0·4047 ha
1 $mile^2$ = 259·0 ha
= 2·59 km^2

Mass

1 ton (US) = 0·9072 ton (metric)
1 cwt (hundredweight, short) = 0·454 quintal = 45·4 kg
1 pound (lb) = 0·454 kilogramme (kg)
1 ounce (oz) = 28·35 gramme (g)

Pressure

1 lb/in^2 (PSI) = 0·06895 bar
= 0·07031 kg/cm^2
= 0·06805 atm
1 bar = 1·013 atm = 1000 cm of water
1 atmosphere = 1·033 kg/cm^2
= 101,325 Nm^{-2}
1 pascal = 1 Nm^{-2}

Flow

1 cusec (ft^3/second) = 0·2832 m^3/sec = 1·0194 hectare-centimetres/hour
= 101·94 m^3/hr
1 million gal (UK)/day = 3·7 acre-ft/day = 4564 m^3/d (cmd)
= 1·867 cusec = 0·0531 m^3/sec
1 cusec for 24 hour = 1·983 acre-ft = 2447 m^2
1 acre-foot = 0·1233 hectare-metre, ha-m = 12·33 hectare-centimetres (ha-cm)
1 acre-inch = 0·01028 hectare-metre, ha-m
= 1·028 hectare-centimetres = 102·8 m^3

1 gal (US)/min = 0·00227 hectare-centimetre/hour
= 0·227 m^3/hr

Force

1 Newton = 10^5 dynes cm^{-2}

Energy and Power

1 erg = 10^{-7} J

1 watt = 1 J sec^{-1}

1 H.P. = 746 watts

Author Index

Subject Index

Numbers in italics refer to the more important area(s) of subject matter